D0341625

René Dubos
Friend of the Good Earth

René Jules Dubos, 1941
Courtesy of The Rockefeller Archive Center

René Dubos

Friend of the Good Earth

Microbiologist, Medical Scientist, Environmentalist

Carol L. Moberg

WASHINGTON, D.C.

Address editorial correspondence to ASM Press, 1752 N St. NW,
Washington, DC 20036-2904, USA

Send orders to ASM Press, P.O. Box 605, Herndon, VA 20172, USA
Phone: (800) 546-2416 or (703) 661-1593
Fax: (703) 661-1501
E-mail: books@asmusa.org
Online: estore.asm.org

Library of Congress Cataloging-in-Publication Data

Moberg, Carol L.
René Dubos, friend of the good earth : microbiologist, medical scientist, environmentalist / Carol L. Moberg.
p. cm.
Includes bibliographical references and index.
ISBN-13: 978-1-55581-340-6
ISBN-10: 1-55581-340-2
1. Dubos, René J. (René Jules), 1901– . 2. Microbiologists—Biography. 3. Ecologists—Biography. I. Title

QR31.D83M63 2005
579'.092—dc22

2005013767

10 9 8 7 6 5 4 3 2 1

Printed in the United States of America

Interior design: Susan Brown Schmidler

∞ The paper used in this publication meets the minimum requirements of the American National Standard for Information Sciences—Permanence of Paper for Printed Library Materials, ANSI 239.48-1192

For lives shared
along the St. Joseph River
and
on The Farm at Henderson Grove

Contents

Preface

On 22 April 1970, nearly twenty million Americans participated in the greatest street demonstration since the end of World War II. Two thousand communities, two thousand colleges, and ten thousand high schools took part. Its *raison d'être* was nothing less than the quality of life on Earth.

The first Earth Day was the largest rally in the country's history. Provoked by years of mounting concern over environmental ravages—from polluted water, air, and food to the infamous Alaskan pipeline, the clear-cutting of national forests, and the proposed damming of the Grand Canyon, among other actual or looming crises—it was the people's way of forcefully calling attention to these insults. Experts were warning of vanishing wilderness, depleted resources, approaching famine, species extinction, population bombs, energy shortages, greenhouse gases, and other eco-horrors. The *CBS Evening News with Walter Cronkite* captured these concerns in the multi-part series asking "Can the World be Saved?"

Among those present that April day was a soft-spoken, French-born biologist whose research had helped revolutionize medical practice. Extrapolating from his knowledge of biology, and of health and disease, he challenged the doomsday view of many of his scientific peers and spearheaded a philosophical and practical approach to ecological health that would transform environmental thinking. "*En garde*, Pessimists!" announced *The New York Times*, "Enter René Dubos."

He called himself a "despairing optimist." In his view, the "real" environmental tragedy was not the destruction of life but its progressive degradation, not death but a worthless existence. The crux of the problem, he argued, was our failure to see ourselves as integral parts of the Earth's ecosystem, an assertion we accept today as axiomatic. His optimism lay in his faith in the resiliency of nature and the ability of human beings to undo the damage they had wrought.

Dubos' popularity during the 1970s was not surprising nor, the *Times* notwithstanding, was it sudden. He had been an authority on the relationship between environment and disease for decades. In 1968, he had written the Pulitzer Prize-winning *So Human an Animal*, which put forth the farsighted view that health is synonymous with ecological well-being. It had been his lifelong credo that a living organism can be understood only through its relationships with everything else.

Dubos' persuasiveness was greatly enhanced by his personal appeal. He engaged audiences with his gentle Gallic accent and avuncular charm. Tall, large boned, and rosy cheeked, with durable white wisps on a balding head, his shy smile was quick to broaden. His attentive blue-green eyes, magnified by thick glasses, radiated inexhaustible curiosity. His large hands punctuated every sentence.

Even more captivating than his manner were his messages. The dramatic speeches and candid concerns heartened audiences who were discouraged by vague political agendas, conflicting data, and confusing expectations from medical technology. With his deep understanding of human potential and limitations, he had the ability to help people understand what they could do—where they lived, worked, and played—about their health, their life, and their environment. He gave the environmental crusade its most famous motto, "Think globally, act locally."

Long before he became famous as an environmental philosopher, Dubos' contributions as a biomedical researcher profoundly altered and enriched medical knowledge and practice. Throughout a long life, many remarkable insights and serendipitous events changed the path of his career several times. The earliest phases focused on soil microbes; he was an agronomist who observed how they decomposed cellulose and then a bacteriologist who used them to develop the first antibiotics. He became a skeptical medical microbiologist who alerted the public to the dangers of depending only on medicine and believing in what he called the mirage of health. As an experimental pathologist he revised the germ theory of disease to show that ubiquitous infection is the rule and overt disease the exception. As an environmental biologist he gathered evidence that led to a field of environmental biomedicine. Finally, as a "humanistic ecologist" he taught that achieving health for ourselves and the Earth depends on making creative adaptations. All the while, as biologist, teacher, consultant to government and private organizations, and public lecturer, he served as a sage *provocateur* who championed the view that the quality of our daily lives is intricately interwoven with the quality of the Earth itself.

This biography tells the story of how a bacteriologist's quest for the mechanisms of disease turned into a philosopher's search for the meaning of health. As his concerns expanded from microbes to humans to society and, ultimately, to the Earth, Dubos became, as much as any individual of the twentieth century, the conscience of health. His ability to think ecologically allowed him to accept and expect nature's changes, grasp its complexities, and fathom paradoxes in health and disease. Along with visionaries Aldo Leopold and Rachel Carson, his essays and lectures reflect his keen observation, philosophical penetration, and eloquence concerning the web of relationships that connects humans to all living things. Echoing Carson's conviction of an "obligation to endure," he elaborated on health as the will or vision to demand what is good for life.

1 Orchestral Relationships and Soil Microbes

The landscape of any farm is the owner's portrait of himself.

ALDO LEOPOLD

FOR FRENCH-BORN RENÉ DUBOS, taking a lesson from a tree meant nothing could be fulfilled in a day. To one who so loved trees, no landscape was more inviting and serene than an avenue of trees. As a child in France, he loved the venerable poplars lining roads, the beeches and oaks traversing parks, and the *allées* of stone pillar-trees along the naves of medieval churches. Nearly every day for over half a century, he strolled under *allées* of sycamores in the Manhattan gardens of The Rockefeller University where he worked as a medical scientist.

Not until he became an American citizen did he appreciate how completely these harmonious leafy tunnels came from human designs with nature. Not until he became a landowner did he experience wildness. And not until he began planting trees did he find himself at peace.

For many reasons, 1946 was the turning point in his life. In that year, when he was forty-five, he bought a ninety-acre abandoned farm in the Hudson Highlands north of New York City. Seven years earlier, in 1939, he had achieved scientific fame by unearthing medicines from cranberry bogs. This was the beginning of many events that transformed him from a notable scientist among peers into a public figure. Yet in the immediate aftermath of his great discovery there were personal disappointments, sorrow, and increased restlessness.

So it was by taking frequent refuge on his land that he gradually established living connections to the soil and found a lasting satisfaction from the hard physical work the land demanded.

Every year after buying the farm, until he was eighty, he planted trees. Taking advantage of the well-watered rocky slopes, he followed ecological principles to harness nature's own powers of recovery. His design for the landscape was to adopt and cultivate individual trees. The fields, returning to wildness, lay entangled in grape vines, poison ivy, briars, and newly sprouted trees. The plan took a great deal of care and experimentation while the hemlocks, maples, and dogwoods he selected grew from seedlings to nurslings to sizable shade trees. With pruning shears, he created a hospitable environment that would allow their fullest shape. He fought off encroaching plants to provide ample air, light, and space. Brush was cleared to free mountain laurels, expose pleasantly shaped rocks, and reveal the natural contours of the hillsides. Trees and plants were arranged to recreate dramatic moods and display hues of green.

These efforts to heal the exhausted farm bestowed a distinct view of conservation that was kindled by affection and continual wonder in the presence of this new wildness. Gone were visions of rural France with its tidy fields, hedgerows, wood lots, pastures, and creeks. Any notions of leafy tunnels were dispelled by the intractable boulders and tentacular tree roots. He was not frustrated by wildness as he worked within its natural constraints, testing textures, colors, and patterns, finding potential, resilience, and renewal while proudly speaking and writing about his improvements on nature.

The land gave new strength to his science and writing. As trees grew, ideas matured and his transition from soil microbiologist to philosopher of earth became more inevitable and more assured. Remarkably, he held no expectation that either his trees, or his ideas, would grow full and beautiful in his lifetime, yet underneath lay the hope that someone might benefit from them in the future.

French Roots and Early Influences

The childhood of René Dubos was shaped by the peaceful, secluded Île de France villages just north of Paris. He was born on 20 February 1901, in Saint-Brice-sous-Forêt, a hamlet of 1,233 inhabitants that was built around a small medieval church with an elegant bell tower. His early youth was spent in the even smaller village of Hénonville with 450 people, at the border of Île de France and Picardy, nestled among gently rolling limestone plains in the Thelle and French Vexin regions. There, the stone houses with kitchen gardens are sheltered by grey walls facing the narrow streets, keeping family life out of public view while social life goes on in the public squares, shops, and cafés. As a village boy wearing a smock, he could push a wheelbarrow of wild carrots for his rabbits along soft country lanes, edged with fragrant hedgerows of hawthorn, and remain within sight and sound of the church bells.

The Île de France has been humanized so long that its second nature cannot always be readily differentiated from its primeval one. It was heavily wooded two thousand years ago, and trees still grow luxuriantly. Most of the primeval forest was cleared during the Middle Ages to create rich farmland, villages, small towns, and industries. This countryside's domesticated fields and managed woodland, along with its low hills, sluggish rivers, and muddy ponds were "nature" to the young Dubos. Its capricious skies alternate between delicate luminosity and threatening darkness. Large fields of wheat, alfalfa, and sugar beet are dotted with small pastures and wood lots. There are also classical parks, once vast forests that had been owned, preserved, and managed by medieval kings and noblemen to satisfy their passion for the hunt. The parks feature wide *allées* of trees that lead to picturesque views of open fields, tidy villages, and Romanesque churches. In continuous use for more than two millennia, the Île de France slowly acquired a tameness and harmony that transcended its primeval endowments. To this day the land supports a great diversity of human settlements.

The name Dubos is common in southwestern France near Bordeaux, where it is pronounced "doo-bow." However, René's grandfather and father were both born in the Île de France, where the family pronounced the final *s*, rhyming their name with the word *dose*. Grandfather Dubos was a house painter in Beaumont-sur-Oise. At thirty-five, he won a national lottery prize, retired, bought three village houses, and settled in one while living on rental income from the other two. His son Georges Alexandre became a butcher in Beaumont. He met his wife, Madeleine Adéline De Bloedt, known as Adéline, while on military service in the small town of Sedan in northeastern France.[1]

Adéline De Bloedt's ancestry is romantic, all the more so because it cannot be documented. As told by René, her father Pierre was abandoned as an infant in the cathedral of Saint Michael and Saint Gudula in Brussels. He was dressed finely and placed in a basket with a note instructing that the baby boy be taken to a farm near Mons along the French border, where money would be provided for his care. The note also warned that no attempt should be made to trace his origin. The infant's name was given as "De Bloedt," Flemish words meaning "of the blood." Pierre De Bloedt eventually settled in Sedan, France, where he worked in a textile factory, and married a humble French woman. Their daughter Adéline, René's mother, received little formal education. She left school at the age of twelve to work as a seamstress.

After marrying Georges Alexandre, Adéline worked in the *boucheries* they ran together, first in Saint-Brice, then in Hénonville. While he managed the *abattoir*, she helped by cutting meat and delivering it in a small horse carriage to neighboring villages. Two more children were born in Hénonville, René's sister, Marie Madeleine, and brother, Francis. All three children helped by feeding the animals and waiting on customers.

Perhaps no event was as devastating and formative in René's early life as the rheumatic fever that first afflicted him when he was eight years old. At the turn of the twentieth century, this was the dread disease of youth, an acute illness that struck early and left its victims with serious heart damage, including valvular lesions that often led to years of invalidism or early death. An American medical textbook of the time, William Osler's *The Principles and Practice of Medicine,* advised physicians that nothing was known about the cause of rheumatic fever although its symptoms of excruciating pain, swollen joints, and shortness of breath were well characterized. More disheartening was the fact that medicines had no effect on the duration or course of the disease. Aspirin was suggested but it often failed to relieve the symptoms and it had no influence on preventing cardiac complications. The only treatment was prolonged bed rest that was made more difficult by an inability to change position without pain, drenching sweats, and otherwise utter prostration. Scientific inquiries into rheumatic fever would not begin to find answers for another twenty years, when many of its mysteries would be unraveled by scientific collaborators of Dubos. Not until antibiotics were available, a discovery in which he played a major role, were methods found to prevent, but still not cure, this disease.

Dubos suffered four bouts of rheumatic fever before he reached thirty-three. The first and most traumatic of these attacks produced irreversible heart damage and affected his health for life. The experience was so vivid that he recalled it on his deathbed, at age eighty, by relating that as a typical husky, vigorous French boy, he was fascinated by bicycle racing and tennis. He dreamed of riding in the Tour de France and playing in the French Open and knew all the current champions by name. One day, he competed in bicycle races with neighboring village boys, winning a race, then returning home in a state of extreme perspiration. Within a few days he developed a sore throat that was followed by severe rheumatic fever. "This resulted in a heart lesion (in the aortic valve), which . . . has always prevented me from engaging in strenuous games or even running. . . . I had to stay in bed almost motionless, suffering from the acute joint pains caused by rheumatic fever, but as much as the pains I now remember the loving care of my mother."[2]

After being indoors for nearly eighteen months, he remembered as "one of the most important events of my life" being allowed to take a short walk outdoors with his mother. "After I had been indoors so long, it appeared to me as an enchanted world. The few people we saw, probably fewer than ten, seemed to me a crowd and made me feel that contact with human beings other than my family was an immensely exciting experience. I then fell deeply in love with the world of things and of people. . . . From then until now, I have known that simply being alive is the greatest blessing we can enjoy."

The intense memories of this early near-death experience and the struggle to recover health reveal much about the inner Dubos. His feelings of anxiety, restlessness, and inadequacy that are common in rheumatic fever patients, plus severe nearsightedness and fear of blindness, were never mentioned. Instead, what he remembered were months of enforced bed rest, ironically, as a pleasant isolation with long hours of reflection and day dreams of *voyages autour de ma chambre*. These voyages were no doubt enhanced by his reading of the profusely illustrated *Le tour de France par deux enfants*, which has introduced French children to all the peoples and regions of their country since it was first published in 1877.

Solitude further sensitized him to his surroundings especially after regaining contacts with the natural world and village life. Always aware that any overexertion could lead to a relapse and death and that the disease would never be cured, he still managed to counter chronic invalidism and fears of hopelessness. Somewhere in those months of loving care from his family he found the willpower to make the most of what was possible, given his disabilities and fanciful childhood dreams. From the onset of illness he showed a special intensity and purpose and presented a lifelong semblance of health. He yearned to succeed and never gave up a goal to become some kind of champion.

He lived with a different kind of health. Painfully shy, he rarely revealed private feelings, especially about health, yet, with a flair for conversation and making others feel comfortable, he made an indelible impression on people. Unable to take part in usual boyhood games and activities, he set his own pace, walking long distances to explore the countryside. Special attractions were walks near Beaumont to Carnelle Forest to rest among its well-preserved druidic dolmen or along the banks of the Oise, where he fished with Grandfather Dubos. Learning to function alone brought out solitary traits of reading and writing that dominated his life. These pastimes enhanced his meditative moods and were the beginning of what he called his freelance spirit.

Dubos loved to tell how childhood stories influenced his later life, attributing an early interest in science to the novels of Jules Verne and an interest in America to the serialized adventures of two early heroes, detective Nick Carter and frontiersman Buffalo Bill, that "created visions of worlds very different from the one I knew."[3] School introduced other worlds. Hénonville had two one-room schoolhouses on its central square, one for boys and the other for girls, both run by a single teacher. As pupils advanced in grade level, they were assigned to teach the younger students. This helped them review their studies and imparted a sense of responsibility for sharing their learning with others. Dubos enjoyed two school exercises, the daily dictations by the teacher from classic world literature and the memorized recitations of similar short texts. Here he discovered places and people far outside the village.

Not surprisingly, many recitations were taken from the fables of La Fontaine. These brief stories featured landscapes and characters of the Île de France countryside, and they fed his imagination as they provided lifelong moral templates. He could still recite them from memory at age eighty. One fable portrayed an old man who was ridiculed by three youths for planting trees that he would not live to see mature, but the old farmer was wiser because he knew others would enjoy the shade from the trees after he was gone. Another fable told of a dying farmer whose sons dug for a promised treasure hoard and found none but were rewarded instead for vigorously tilling the soil, improving the land, and producing larger crops that brought them great fortune. Later in life, he used fables like these to simplify complex environmental issues and create picturesque messages.

When René was thirteen, the family moved to Paris. His father opened a butcher shop at 46 rue Alexandre Dumas near the Place de la Nation in a neighborhood of many small workshops. The shop was closed in August 1915 when he was called to active duty in World War I, although his service was seriously limited due to chronic ear ailments and obesity. He reopened the shop in 1919 on his return but died a few days later following an ear operation. Adéline was left to raise three children and run a small *épicerie* in the Paris suburb of Livry-Gargan.[4] The teenage René enjoyed helping his mother run the store, recalling "that the most useful part of my education, from the human point of view, was to watch my mother converting the sale of a lamb chop into a pleasant social event."

Through the generosity of one customer, Auguste Panon, a copyist of French impressionist painters, Dubos learned to paint. Although no paintings survive, he must have been adept for he practiced this hobby for several years. As late as 1930, he and several French friends spent their summer vacations in Auribeau, Provence, where they painted landscapes and copied old masters.

René attended high school on a scholarship at Collège Chaptal in Paris. Despite a second bout of rheumatic fever at the outset of these studies, he maintained an excellent academic record. The curriculum emphasized literature, history, and German and English rather than Greek or Latin and provided his first exposure to science.

While at Collège Chaptal, a strong interest in history led René to Hippolyte Taine's *La Fontaine et ses fables* (1875) and introduced him to the molding force of the environment on historic events. Taine argued that if La Fontaine had been born in German forests, Sudanese deserts, or Mediterranean countries, the fables would have been entirely different. La Fontaine's poetic temperament had been shaped by the landscape, climate, and people of the Île de France, and his lessons about good and bad behavior revealed regional solutions to universal problems that mirrored all of human nature. In many ways, Taine's thesis influenced Dubos' scientific and social beliefs that man makes himself while shaping his environment.

With his father's death, Dubos wanted to quit school to work full time with his mother, but she was determined that her son not abandon his education. He remembered how the two of them studied the special pages in *Le petit Larousse* on the *grandes écoles* to plan for the next level of schooling. In 1919, he received a baccalaureate for his secondary school studies and was preparing for an entrance exam to L'École de Physique et Chimie. Then, a series of chance events altered these plans and initiated a different career. He suffered a third bout of rheumatic fever and on recovering passed an exam to the one school still open for enrollment that year. Thus, he entered the Institut National Agronomique, known familiarly as L'Agro, whose focus was training students to work as agricultural experts rather than researchers or scientific farmers. "It teaches you to have a scientific mind about agriculture," he said, "It's for the head, not the hands."[5]

He did well in all the school's technical courses except microbiology, which he found intensely boring because it dealt solely with taxonomy. Chemistry was so distasteful he told his mother he would never again enter a laboratory. The courses in which he excelled were *économie rurale* and *technologie agricole*, which dealt with a farmer's decisions about land to cultivate and the relation of soil fertility to climate. In 1921, after completing a second year as twenty-third in a class of sixty-one students, Dubos received a "college" diploma as *Ingénieur* from L'Agro.

The following year, he received another diploma as *Ingénieur* from the Institut National d'Agronomie Coloniale in Nogent-sur-Marne, where he had a scholarship to study for a career as colonial administrator of technical affairs in Indochina. In 1922, he entered an officers training school in the French Army. Even though his military card indicated a heart murmur, he engaged in demanding maneuvers and drills. A few months later, after overexerting himself, he was discharged because of valvular insufficiency and thus became ineligible for the post in Indochina. This new disappointment left him without plans or prospects for employment. At this point, he developed a duodenal ulcer, perhaps brought on by taking aspirin for his rheumatic condition, and this added another burden to his lifelong afflictions.

Paris exerted special influences on Dubos' teenage and young adult years. The city's great diversity of public areas, river banks, and street corners helped him discover what he wanted to do and to become. During hours spent on park benches observing human nature and pondering various lifestyles, he was impressed by people of all ages who were engaged in play, love, deep thought, or animated conversation. He had become aware of an immense diversity of people from fables, history, and limited personal experiences, but he despaired that many aspects of Parisian life were unavailable to him if only for economic reasons. While distancing himself from opulent lifestyles observed in Parc Monceau near Collège Chaptal, he was attracted to the bohemian lifestyles of students reading and debating in the Jardin du Luxembourg near L'Agro. Choosing a life of intellectual rather than economic adventures seemed settled at this stage. Even after reaching America, he bragged in letters home about his *vie bohème* in New York.

Several months after being disqualified from colonial service, Dubos found a job in Rome that took advantage of his agricultural expertise. In February 1923, he joined the staff of the International Institute of Agriculture, established in 1905 to serve as a worldwide information service. Dubos worked in its Bureau of Agricultural Intelligence and Plant Diseases, which reported current developments in the science and techniques of farming, agricultural industries, livestock improvement, rural economics, and plant diseases. He was one of six editors of the *International Review of the Science and Practice of Agriculture*. This quarterly journal, published in English, French, Italian, and Spanish, featured abstracts of international agricultural reports from books and journals. During an eighteen-month tenure as editor, his byline appeared on nearly 150 analytical summaries, or an average of twenty-five an issue. Ranging in length from a paragraph to several pages, these summaries were the first publications under his name.

Some clues in these abstracts suggest an emerging author. Even though this position is where he learned to speak English and Italian, Dubos certainly wrote these pieces in French. Almost eighty percent of them were abstracted from journals originally published in French on *agronomie coloniale* in locales as diverse as Africa, India, the Philippines, Vietnam, and Poland; the others were abstracted from English and German publications. The *International Review*'s English version of his abstracts contains well-crafted sentences linked by a typical French rhetorical structure, suggesting these were translated by someone else, since they were unlike the English he used in his first American publications.

The abstracts are striking for their precise scientific and technical explanations. They show familiarity with mechanical operations of farm implements and agricultural production techniques, including topics on wine, cider, potatoes, sugar beets, tropical oils, and suitable plows and harvesters for developing countries. In contrast, he wrote a single report on microorganisms. A few topics anticipate his future interests in land management, alternative resources for fuel (straw, wind, palm oil), and significance of the soil. By nature, the abstracts lack editorial comments, yet in his final appearance as editor, he dared offer a personal and somewhat prophetic observation about ecological phenomena in soil processes: carbon dioxide production of the soil, he announced, is affected by relationships among cultivation methods, fertilizers, and the nature of the soil itself.[6]

By the time this abstract appeared in print, Dubos had encountered Sergei Winogradsky and soil bacteriology and had landed at an Agricultural Experiment Station in the United States.

Approaching a Career in Soil Science

While a science devoted to soil is scarcely a hundred years old, its beginnings were to touch René Dubos in many ways. By the 1920s, some scientists were advancing the novel idea that soil is a natural body, like other natural systems, and deserved its own scientific study. Even defining the word *soil* at the time was problematical. According to early soil scientist Hans Jenny, every soil is individual. The "processes in the soil are orchestral. They deal not with the infinitesimals of chemistry or physics or biology alone but with interacting realms whose concert is the soil's life."[7]

Dubos' immediate touchstone to this view was one of the founders of soil microbiology, Sergei Winogradsky, the Russian botanist and chemist whose studies of soil microorganisms began at the end of the nineteenth century.[8] His work came twenty-five years after Louis Pasteur speculated on their role in the decomposition of organic matter and stimulated the search for and isolation of microbes involved in animal and plant diseases. Isolating the variety of microbes from soil, however, presented a more difficult situation. Winogradsky in Russia and Martinus W. Beijerink in The Netherlands made decisive contributions by isolating the nitrifying, nitrogen-fixing, and sulfur-oxidizing bacteria from soil. They also showed the major role of microorganisms in converting organic matter into inorganic matter that can be used for growth of higher plants and animals. Their work established fundamental biological processes in soil fertility with far-reaching agricultural applications. Despite their discoveries, further scientific development of an agricultural microbiology did not expand until another quarter century passed.

Following political upheaval in Russia after World War I and the 1917 Revolution, Winogradsky was forced to leave his homeland forever. His early studies in Russia were so well known that Emile Roux, director of the Pasteur Institute, soon invited Winogradsky to organize a division of agricultural bacteriology in France. He arrived in mid-career to join the Institute. From 1922 until his death in 1953, Winogradsky lived almost as a recluse on a small estate twenty miles outside Paris in Brie-Comte-Robert, where he continued to work and publish on problems of soil organisms.

Two years after Winogradsky settled in France, Dubos almost but never quite met the man he called "the deepest influence on my intellectual life."[9] This happened on two occasions, both during the spring of 1924 in Rome.

Dubos first encountered Winogradsky while searching for articles to abstract in the *International Review*. He later recounted the excitement of finding Winogradsky's call for a "dynamic microbiology" devoted to soil. The article appeared in the February issue of a semi-popular French journal *Chimie et Industrie* and was not one he abstracted for publication.[10] However, its emphasis on a *dynamic* science probably changed his mind about microbiology being boring. Dubos claimed Winogradsky's ecological approach was so influential that he decided to "become a bacteriologist."[11] Winogradsky had revealed that soil is a living organism with endless webs of relationships in which microorganisms are the indispensable links, controlling both growth and decay to foster life in the soil. What Dubos could not foresee at the time was how his future work and philosophy would be so firmly guided by the lessons and principles set forth in this article. In fact, they characterize his approach to biological research.

During his nitrogen-fixation studies, Winogradsky wrote, "I had determined that the failures of my predecessors were due to the utilization of gelatinized media which are so widely employed at present for the isolation and cultivation of microbes. The nitrifying organism refuses to grow on such media." Winogradsky was criticizing the "elective" or agar plate method developed by Robert Koch to obtain pure specimens of microbes. This method is a mainstay of the practitioners of the germ theory of disease as a means to implicate a specific microbe as the cause of a disease, and it continues to be used for the diagnosis, management, and study of infectious diseases.[12]

Winogradsky maintained that the study of isolated bacteria gives a false picture, because it is their behavior in groups, or their social dimension, that provides the key to understanding them. He called artificial (pure) cultures a horticultural or "green house" science: when microbes are removed from their natural habitats, free from competing influences of other microbes, they are likely to manifest different activities. Each organism in a pure culture acquires a degree of specialization and responds in a particular manner, so its activities may be either under-, over-, or unexpressed. For example, depending on nutrients in the media, rare organisms may show enhanced growth while common ones are repressed.

In the article, Winogradsky argued that soil processes must be studied as a whole and under natural conditions: "soil is a living world saturated with a teeming mass of microscopic beings, having a variety defying all imagination." It is not the "immanent qualities" of microbes that determine their role, he explained, "but the battle of cells with all these pressures which is the principal apparatus for controlling and dispersing microbial activity." The number and kind of microbes are determined by physical and chemical characteristics of the soil, and they in turn condition the properties of soil and its fertility.[13]

Winogradsky therefore promoted direct methods for studying soil microbes. The direct or enrichment culture method involves filling glass cylinders or tumblers with fresh garden or field soil samples and using them as the culture media. The ecological concept behind this method assumes that by adding or feeding these miniature ecosystems with specific organic substances, which act as energy sources, it is possible to isolate the microorganisms utilizing these sources and working in these environments. Winogradsky admitted inherent difficulties with studying soil. One is its opaqueness, which makes it difficult to analyze teeming microbes under the microscope. Another problem is that the chemical composition of a *paysage microbien* varies enormously from one soil to another. Finally, soil is a poor culture medium.[14]

Using direct techniques, Winogradsky made three important discoveries: each type of soil contains an indigenous biota, each sample of soil harbors innumerable types of microbes, and many soil microorganisms could not be cultivated by any other means. As a result, he claimed that soil is an aggregate of diverse, nonreducible components and functions as more than the sum of its discrete parts and processes. At any given moment, most soil microbes are dormant, so it is only by studying these static and dynamic phases that one could speak of a science of soil microbiology. Dubos would soon appreciate many of these orchestral relationships while doing experiments for his doctoral research.

Dubos' second encounter with Winogradsky took place in May 1924, just a few weeks after reading the article in *Chimie et Industrie*. The International Institute of Agriculture was conducting the Fourth International Conference of Pedology. (Pedology was an early name for soil science.) Scientists who attended the congress came from two dozen countries and sixty scientific institutes. Among the general lectures given by illustrious members of agricultural science was one by Winogradsky on the direct methods of studying soil. Another lecture was by Jacob Lipman, dean of the College of Agriculture and director of the New Jersey Agricultural Experiment Station at Rutgers University, who spoke on the relation of soil science to the fertilizer industry.

At the concluding meeting, the International Society of Soil Science was founded. The International Institute of Agriculture in Rome was selected to become the center of activities of all branches of soil science. Lipman was elected its first president and appointed organizer of the next soil congress to be held in the United States in 1927.

Dubos served as an assistant at the Rome conference but neither heard nor met Winogradsky. "I did not exchange a single word with the man who has the most profound influence on my life since I reached age 23," although from some twenty feet away he watched the "tall, imposing, and very distinguished" scientist walk through a large crowd.[15] Why he never met Winogradsky during the following thirty years is puzzling. It would have been easy to journey the few miles outside Paris to meet him during his many visits to stay with family in Paris and to call on scientists at the Pasteur Institute. Selman Waksman, his professor at Rutgers, who frequently corresponded and visited with Winogradsky, could have introduced the two or suggested a visit.

The conference nevertheless put Dubos in touch with many agronomists. His duties at the conference were to guide visiting scientists around Rome. These were not the pedological excursions to examine various soils surrounding the city, but rather sightseeing excursions within Rome itself. This is how he met American delegates who urged him to study in the United States. One was Asher Hobson, from the University of Wisconsin, who reportedly told him, "I know a winner when I see one." The other was Jacob Lipman, who mentioned a possible position at the New Jersey Agricultural Experiment Station. This is also where he first met Selman Waksman, a soil microbiologist from the Experiment Station. After the conference, Waksman traveled throughout Europe to assess the state of soil microbiology in various laboratories and learned from a young German scientist that "to call oneself a soil bacteriologist at the present time is to ruin entirely one's chances of advancement in the future." Waksman concluded that soil science was "in its infancy and one may rightfully ask whether such a science exists at all at the present time."[16]

Further encouraged by comments from visiting Americans, Dubos took a summer course in bacteriology and saved money for his voyage to the United States by translating books on forestry and agriculture into French. From that moment, he lived an immigrant's dream to have adventures in

America. The urge that brought him to the New World in 1924 was not motivated by the desire to escape from a hostile environment. Science was only one incentive. His childhood reading about adventures of American cowboys and immense solitudes of a wild continent also nourished dreams of freedom, abundance, and unlimited possibilities. Drawn like countless, poor Europeans with similar utopian visions, he took a measure of strength and pride that a stranger in America was not asked, "What is he?" but instead "What can he do?" On leaving France, he had no immediate plans for the future, no job in view, not even a student visa—just a grand sense of expectancy.

Then, in one of the many coincidences of Dubos' life, when Waksman returned to the United States that September, the two were fellow passengers on the *Rochambeau* from Le Havre. As Waksman recalled, "The sea voyage gave us ample opportunity to compare educational systems in America and in Europe, and to discuss the beginnings of great scientific developments in the United States."[17] Clearly, during this voyage, Waksman invited him to study in America. After docking at Ellis Island, New York, on 30 September 1924, Dubos wrote into the ship's passenger record that his occupation was "civil engineer" and that his destination was the College of Agriculture at Rutgers University in New Jersey.

Entering the New Field of Soil Microbiology

Selman Waksman, like Winogradsky and Lipman, was a Russian emigré. He came to New Jersey in 1910, lived on a farm with a cousin, and then trained as a soil bacteriologist under Lipman at Rutgers by recording the abundance and kind of microbial populations in various levels of soil at the school's farm. After he earned a Ph.D. at the University of California in 1918 in biochemistry of soil microorganisms, Lipman appointed him to develop the field of soil science in the New Jersey Agricultural Experiment Station at Rutgers. Waksman spent essentially his entire scientific life there studying the soil-dwelling actinomycetes, the fungus-like bacteria that grow as branching filaments. According to Hubert Lechevalier, a former Waksman student, seventy-seven students received advanced degrees in soil microbiology under Waksman. Some of his major research interests were the role of microbes in decomposition of organic matter, the development of humus, the production of enzymes, the formation and utilization of peat, and eventually the microbial antagonisms and antibiotics produced by the actinomycetes.[18]

Waksman further inspired Dubos' ecological approaches to bacteriology. There is little evidence to indicate that Dubos or Waksman were influenced by a discipline of ecology that was just emerging in America during the 1920s.

The words *ecology* and *human ecology* did not appear in Dubos' writings until the 1960s, although he used the adjective *ecological* sparingly. As used in this biography, *ecology* and *ecological* refer to an attitude or matter of conscience—seeing things in perspective, making connections, taking a multidisciplinary approach to problems—and not to a predictive science.[19]

Dubos remembered his professor as having the "most comprehensive knowledge and understanding of the complexity of biological phenomena, and moreover, of the interrelationship between biochemical processes and biological processes." Compared with academic courses he took on the Rutgers campus, the scientific training at the Experiment Station was much broader. Waksman taught developmental, environmental, and historical aspects of bacteriology. His lectures featured contributions of those European scientists he visited, bringing "a living expression of what soil microbiology was in the laboratories of those who were developing it." Dubos was intrigued by Waksman's analysis of the two men who had been more than anybody else influential in shaping soil microbiology, comparing Winogradsky, the classicist who quantified laboratory phenomena, with Beijerink, the romanticist who preferred natural history and ill-defined problems.[20]

At the same time, Dubos was critical of Waksman, contrasting his casual style with the kind of rigorous intellectual logic that students had to produce in French schools. The American courses were too easy, he said, the atmosphere too relaxed, the presentation of science too sloppy, and no real laboratory training was associated with Waksman's own courses. What bacteriological skills he mastered came from being a teaching assistant for another Rutgers professor, where his duties involved preparing all materials and culture media for the course. Despite disappointment in Waksman's operational skills, Dubos claimed it was through such perceptive and comprehensive views that he quickly "developed an ecological concept of microbiological processes at a time when everybody else was becoming more and more specialized."[21]

Waksman regarded Dubos as the "youngest, certainly the quietest" of his students, observing that "when he had something to contribute it was well thought out and logically presented."[22] The Experiment Station provided a stimulating international atmosphere, similar to that of the Institute of Agriculture in Rome. From 1924 to 1927, the graduate students in Waksman's group laboratory included two Danes, two Russians, a few Americans, a Brazilian, and an Englishman, all of whom Dubos thought were better trained technically than he was.

The young Frenchman could not take part in many extracurricular activities for he had to work part time to earn extra money. Among the various jobs, from animal caretaker at nearby Johnson & Johnson, to tutoring and baby-sitting the children of Lipman and Waksman, to washing laboratory glassware on holidays, one proved valuable for entry into medical research. Fred Beaudette, an avian pathologist at the Experiment Station, learned of Dubos' language skills and paid him to translate European papers on poultry diseases into English. These writing tasks helped him practice scientific English and exposed him to pathology for the first time. The work also introduced him to an ecological science, since Beaudette's research focused on the effect of poultry on soil microorganisms and the soil's effect on the health of the flock.

In 1924 the Waksman laboratory was trying to measure chemical effects of microbial activities to see whether they could serve as indexes of soil fertility, a topic of interest to Director Lipman for years. Waksman's procedure was to involve students in some aspect of his immediate research rather than to let them develop a theme of their own. As part of his master's research, Dubos was given the task to assess the activity of catalase, an enzyme that decomposes hydrogen peroxide into oxygen and water. His first publication reporting these experiments appeared as number ten in a series of papers from the laboratory on soil fertility.[23]

Earlier papers in this series had been criticized by Winogradsky in a letter to Waksman dated 5 November 1924, just a month after Dubos joined his laboratory. Winogradsky wrote, "The question whether much can still be expected of these methods will have to be answered in the negative."[24] It is unknown whether Waksman apprised Dubos of Winogradsky's criticism. However, in keeping with a developing independent manner of research, Dubos' approach to the project was more influenced by Winogradsky than Waksman.

In the introduction to his 1926 master's thesis, Dubos dared express negative feelings that "A priori, the peroxide test is artificial and empirical." He concluded that release of oxygen from hydrogen peroxide was due more to nonbacterial actions and to "the summation of too complex phenomena to allow" its use as an index of soil fertility. After dismissing Waksman's factorial approach as lacking integration, he suggested another way to make "a synthesis of the information acquired and based on a thorough knowledge of the chemical, physical and climatic soil conditions." These comments attest to an aspect of Dubos' scientific temperament that did not accept just one side or one solution to any problem. His observations, although tactfully stated, were bold and self-confident and indicated there were multiple complications of the problem at hand.[25]

In the summer of 1925, Dubos and Charles Edward Skinner, his first friend in America, drove to the Pacific Coast. Skinner had just received his

Ph.D. in Waksman's department and was going to visit his parents in Washington State before spending a postdoctoral year in Rothamsted Experimental Station in England with its leading soil scientist, Edward J. Russell. Crossing the continent during midsummer on mostly graded or graveled roads west of Illinois was a strenuous adventure, but Dubos said, "I felt completely rewarded for our tribulations when we entered Nebraska and Wyoming—Buffalo Bill's holy hunting grounds. The Far West was equal to everything I had imagined." His mood changed when they reached the Snake River Valley in Idaho where the tall poplars and green valleys reminded him of France, and the "experiences in overnight cabins and inexpensive restaurants revealed a human world far different from what I had known." Above all, his first view of the Pacific Ocean from an elevated point in an evergreen forest near the Columbia River "was like the completion of my discovery of America." During their return to New York, the two adventurers made a pilgrimage to Lookout Mountain near Denver, where Buffalo Bill is buried. More than fifty years after this trip, Dubos told environmentalist Tom McCall, governor of Oregon, this was "one of the great events of my life because for the first time I had a feeling that I had made contact with the whole world."[26]

The long hours spent with Skinner on this trip clearly influenced the topic of Dubos' doctoral research, cellulose decomposition. At the time, a major unknown was how organic matter like cellulose is continuously recycled from dead plants into soil. In 1923, Russell announced, "We are at present ignorant as to which organisms are most efficient in decomposing cellulose in the soil under field conditions, or what are the conditions best suited to their activity." He added that the subject "offers one of the most promising fields of research in soil bacteriology."[27] Among the controversies involving this topic, one predominated. V.L. Omeliansky, Winogradsky's former assistant in Russia, had reported in 1913 that decomposition under natural soil conditions was limited to the anaerobic bacteria.[28] In contrast, American scientists about that time claimed decomposition was an aerobic process.[29] Waksman and Skinner were exploring yet another hypothesis, that decomposition was controlled by the availability of nitrogen rather than by specific microbial biota and that fungi were the predominant organisms involved in this process. Citing Winogradsky's direct culture methods as presenting "various difficulties," Skinner based his research on Waksman's 1924 egg albumin agar plate methods.[30]

Dubos elected instead to use for his thesis work the approach advocated by Winogradsky to study decomposition in "soil as a whole." He announced it was an investigation of "ecological" phenomena involved in cellulose decomposition in which attention was devoted "chiefly to the conditions affecting the growth of the cellulose decomposing bacteria." This required a test system where microbes could function by interacting with their natural environments. Notably, these experiments produced results far different from those Skinner and Waksman got using pure cultures of microorganisms growing on cellulose. Echoing Winogradsky, Dubos noted that such an isolation of any organism "is no proof that the organism takes an important part in this process under natural conditions."[31]

Adapting Winogradsky's direct method, he evaluated many soil samples to determine which microbes decomposed cellulose under a variety of "normal" conditions. The experiments tested such contrasting factors as acidity/alkalinity, moisture/dryness, and oxygen/nitrogen as well as environmental variables including nutrients, available space, climate, chemical composition, other types of cell populations, and age—all of which can produce either associative or antagonistic processes. This was followed by evaluating the extent of bacterial activities and isolating active microbes using silica gel methods. The technique produced abundant organisms, many that were previously unknown, but unfortunately Dubos could not relate their number to the inoculum he used. As Winogradsky had predicted, the frustration of not being able to quantify results simply underscored the challenge and difficulty of working with soil. Appropriately, this section of his thesis contains the only stated reference to Winogradsky.[32]

Nonetheless, Dubos' interest in microorganisms was aroused. Waksman remembered arriving at the laboratory one morning to find his student excitedly watching cellulose being disintegrated by a bacterial culture he had recently isolated. "His eyes were shining. His face was intense. As he watched the disintegrated cellulose fibers and the yellow slime produced along the bacterial streak, he was dreaming of processes of a new and little known group of microbes that played its part in the cycle of life in nature. I then recognized that he was on his way toward a real scientific future."[33]

As a result of this research, Dubos revealed a dynamic ecosystem in the soil and made two discoveries: cellulose was decomposed by a number of different organisms and, more important, the type of organism responsible varied with the soil type. In particular, fungi were active in acid soil, actinomycetes in dry soil, and bacteria in moist soil. Some kinds of microbes thrived exclusively on the cellulose while others existed in a resting or latent form, a condition that proved important in his later discovery of antibiotics. Moreover, the chemicals produced during the decomposition process varied with both microbes and soil conditions. Overturning claims made by Waksman and Skinner, Dubos reported that fungi were not predominant and that nitrogen was in fact "insignificant as far as it tends to determine the nature of the cell decomposing flora . . . [and] does not affect appreciably the decomposition of the cellulose." Still, criticizing the value of his research, he concluded that when dealing with such a complex system, "it is probably impossible to reach any definite and absolutely general conclusions." Nevertheless, by linking a concert of microbial relationships to both destruction and creation in the soil, Dubos embraced the orchestral nature of ecosystems.[34]

Three publications resulted from these doctoral studies. Significantly, an abstract by Waksman and Dubos on various microbes involved in decomposition was presented by Winogradsky to the French Académie des Sciences in November 1927.[35] Two months later, Dubos was sole author of another article on the varying influences of soil conditions. This very early report shows two prominent facets of his writing: an allusion to an underlying intellectual concept behind the science and the emergence of his key belief that the growth of a living organism is intimately related to its environment. At the end of his life, Dubos noted with shock that this article had appeared in a relatively new journal called *Ecology* and how totally unaware he was at the time of a scientific field by this name.[36] The third paper reported his isolation of five previously unknown soil microbes.[37]

Winogradsky was keenly aware of Dubos' work at Rutgers and cited all three papers in the summary of his 1929 monograph on soil microbiology. Commenting on the difficulty of performing these studies, Winogradsky granted it was practically impossible to chemically define a continuously changing environment in the soil. The critical issue, he wrote, is to recognize that natural degradation "concerns not only a purely chemical effect, but also a biological adaptation, comparable to the situation of parasitism, which must enter into play in order to attack insoluble organic matter having a complex structure." His conclusion emphasized Dubos' discovery that environmental factors, or his term "secondary conditions," change the microbial balance in various soils. "It is not only the quantity of the cellulose decomposed," Winogradsky wrote, "but especially the quality of the action which commands attention."[38]

Several qualities from this early work characterize Dubos' approach to future research. Among the lessons derived from soil microorganisms was a grasp of their complexities, a sense of Earth as a resilient organism. There was confirmation of Winogradsky's observation that environments determine microbial responses in ways that laboratory tests cannot reproduce natural processes. Moreover, interacting microbes were found to differ qualitatively from isolated microbes. He displayed independence by trying novel methods and challenging the observations of others. An important clue to his thinking, stated in the thesis, is that the research had not so much resolved questions as implicated other integral factors.

By the age of twenty-six, Dubos was scientifically directed toward the interplay between environmental factors and living things. These were seeds of ecological ideas that would be developed during the rest of his life. One focused on how natural ecosystems embrace all types of individual organisms, in which each one carries out a certain behavior. Another emphasized that living processes are not isolated events but are activities carried out in relation to all surrounding forms of life and matter. Health—whether it pertained to soil fertility, or, later, to human disease or pollution on Earth—relates to the *quality* of biological adaptations taking place in the environment.

Developing these ideas in the next stage of his career in medical research quickly led to his first major discovery. He then took advantage of the specialized behavior of microbes and discovered bacteria that would kill other bacteria—the major step in launching the antibiotic era.

2 Domesticating Microbes

O how can it be that the ground itself does not sicken? . . . . Though probably every spear of grass rises out of what was once a catching disease.

WALT WHITMAN, *The Compost*

THE ROCKEFELLER INSTITUTE FOR MEDICAL RESEARCH, founded in 1901, was the first organization in America devoted to scientific medicine. In Europe, medical research was advancing at a number of universities and at private institutes, especially those created for Louis Pasteur in France and Robert Koch in Germany. These two scientists had discovered microbes can cause many diseases, and this led to immediate practical applications for controlling disease. Their success created an aura of expectancy about the scientific basis of medicine, yet at that time, there were so few research laboratories in the United States that physicians aspiring to careers in research trained in European centers.

The Institute was not conceived by physicians or scientists but by an ordained Baptist minister, Frederick T. Gates, who was an adviser to John D. Rockefeller in matters of philanthropy. In 1897, Gates read Osler's textbook *The Principles and Practice of Medicine* to understand why "the best medical practice did not, and did not pretend to, cure more than four or five diseases."[1] He quickly realized that causes of diseases as well as remedies could be found if facilities existed for intensive research. In the same year, he proposed to Rockefeller that he found a new institution for medical research. Mr. Rockefeller's careful preparations accelerated in 1901, when, after his first grandson died from scarlet fever, he learned from the child's physicians that they knew neither what caused the disease nor how to treat it. The Institute was incorporated in June, free of university, government, or commercial control.

The Institute served as a "workshop of science" where scholars could develop knowledge of the nature and cause of disease as well as methods for prevention and treatment. Gates advised that medicine could hardly hope to become a science until "qualified men could give themselves to uninterrupted study and investigation, on ample salary, entirely independent of practice." Thus, the Institute was devoted to scientific culture and its staff was freed from clinical duties, teaching, examinations, and administrative responsibilities. The laboratories were organized around a few talented investigators who were given autonomy and flexibility to direct their own fields of study. When the Institute opened a hospital devoted to clinical research in 1910, its physician-scientists studied patients suffering from a few major diseases for which medicine had no answers: lobar pneumonia, syphilis, rheumatic fever, and heart and kidney diseases.

Initially, microbiology and pathology were the predominant sciences at the Institute, reflecting the impressive discoveries made in European institutions. This focus changed rapidly to a more biochemical outlook, and bacteriologists were then joined by chemists, physiologists, and biophysicists. Simon Flexner, the Institute's first director, had trained in pathology and bacteriology at Johns Hopkins University and then in physiological chemistry in Europe. He strongly supported a mission of applying physical chemistry to biological processes. One of the first scientists he recruited was a German-born physiologist, Jacques Loeb, who arrived in 1910. Calling medical science a "contradiction in terms," Loeb advocated the study of greatly simplified biological systems by methods of chemistry and physics. He also believed experiments should be reduced to analyses of chemical substances that could be separated from biological materials. The philosophy expounded in his book *The Mechanistic Conception of Life* (1912), according to Dubos, "put an indelible stamp on the scientific approach to medical research at The Rockefeller Institute."[2]

It may seem extraordinary, even peculiar, how a soil microbiologist could think of joining Rockefeller's research hospital. At the time, Loeb's influence appealed to many young scientists, including Dubos, who initially believed working and thinking in such an atmosphere was to study "those problems dealing with the most fundamental biochemical phenomena of life."[3] The Institute's structure also fostered continuous daily contacts among its fifteen laboratories and these led to numerous collaborations. Biologists could consider problems in chemical terms, and chemists and biophysicists were encouraged to apply their knowledge and techniques to medical problems. Even among the Hospital staff, many were not physicians but, like Dubos, held Ph.D.s in chemistry, physiology, or microbiology.

Another fortuitous event in Dubos' life came as he was graduating from Rutgers with few plans except to move out of the practical aspects of soil science. His application for a National Research Council Fellowship in the Biological Sciences was withdrawn in May 1927 because he was not an American citizen. However, the secretary who informed him of his ineligibility also sent a note recommending that he consult with a fellow Frenchman, Alexis Carrel, at The Rockefeller Institute. Medical research seemed out of the question but he decided to contact Carrel.

The Institute in 1927 was a sequestered, unpretentious, and self-sufficient community. Located on Manhattan's Upper East Side, it was then far from New York's noisy, bustling centers of activity. When Dubos arrived at York Avenue and Sixty-sixth Street, he entered through an imposing gate in the iron fence surrounding the Institute's fifteen acres, climbed the hill lined with an *allée* of young sycamore trees, and reached the few beige brick buildings on a rocky cliff forty feet above the East River. It was a walk he would take almost every day for the next fifty years, and he never wavered from his first impression, "I know this is the place."

Carrel was an experimental surgeon whose current interest was the chemical nature of malignant cells. He had won a Nobel Prize in 1912 for his mastery of suturing blood vessels, a technique at the foundation of heart surgery and organ transplantation. Dubos had never heard of Carrel and knew nothing of his work. This famous, busy scientist welcomed the young Ph.D. with the advice never to accept a teaching appointment in a small school since it would lead nowhere. Although he could offer no help in finding a position in microbiology, Carrel took him to the Institute's lunch room. Whether by chance or prearrangement, they sat at a table with Oswald Avery, a physician-bacteriologist in the Hospital.

Avery was a small, slender man whose charming, courteous, and conservative ways reminded Dubos of a "buttoned-up petit bourgeois." Everything about his person was "low key like the inconspicuous buildings in which he lived and worked." Avery's scientific ideal, following the model of Loeb, was to define chemically the substances involved in biological problems. As a result, his persistence as an investigator, continued scientific creativity, and lifelong focus on the pneumococcus bacterium led him in 1944 to the discovery that DNA is the genetic material, now considered one of the most important discoveries of the twentieth century.

Dubos often called the serendipitous meeting with Avery on 2 June 1927 "*the* most important event in my life." It was the beginning of a creative scientific association and close friendship between the persistent teacher and eager student. After lunch, Avery took Dubos to his laboratory and asked for details about his experiments on microbial decomposition of cellulose. He then proceeded to link the young man's work to a long-standing research problem. Years before, he had discovered that a capsule made of polysaccharide covers the pneumococcus. Since that time, all the experiments trying to decompose this substance had been futile.

Opening a desk drawer, Avery removed a small tube containing a purified powder form of the capsular material. He waved it in front of Dubos with the enticing remark that if something could be found to dissolve the capsule, and—importantly—if this substance was also harmless in the body, then something could be done to cure the most deadly form of pneumonia. Persuaded by Avery's intensity and propelled by his own sense of adventure, Dubos claimed he could find such a substance. Brash and confident, he believed this problem could be solved because the capsular material was a kind of cellulose. Avery was impressed by this soil scientist and introduced him to Rufus Cole, director of the Rockefeller Hospital.

That same day Dubos formally applied for work at the Institute. Cole told Flexner that before appointing Dubos he wanted to hear from Waksman to "make quite certain that he has good technical skill and ability. It seems quite certain that on the theoretical side he is all right and has plenty of enthusiasm." On 7 June, Waksman wrote to Avery, recommending his student "most heartily" and commenting that his interest in anaerobic processes had familiarized him with both current ideas and methods on the subject. Because Dubos had spent "a good part of his time in preparatory courses such as biochemistry and physical chemistry, he has succeeded in carrying through a very fine piece of work on the decomposition of cellulose." Waksman added "he is a very keen student and is capable of grasping a new idea and developing it further very readily."[4]

Based on Cole's recommendation, on 17 June the Institute's Executive Committee approved Dubos for a one-year appointment as fellow at $1800. This lowest rank on the scientific staff enabled the Institute to engage talented people who could test novel ideas about specific scientific problems. The position was considered temporary, either because the candidate was on trial or because a member wanted "to augment for a year or two the staff of a laboratory in which some very active development is taking place."[5] Four days later, on 21 June, Dubos accepted the position in a letter handwritten on stationery of the Willard Hotel in Washington, D.C. He thanked Flexner for "the honor conferred upon me and beg you to believe that I will do my best to show myself worthy of it."

During the next few weeks, Dubos participated in the First International Congress of Soil Microbiology, organized by Lipman of the Experiment Station.

It was opened in Washington, D.C., by President Calvin Coolidge, who grew up in a Vermont farming community and was personally interested in the progress of scientific agriculture. Winogradsky did not accept an invitation to journey to America, so Waksman presented his paper on the direct methods of studying soil. Dubos, in what was certainly his first public presentation of science, reported on his studies of cellulose decomposition during a session led by Edward J. Russell, the English soil scientist.

Following the meetings, two hundred of the seven hundred delegates left on a transcontinental excursion by chartered train. As a delegate of France, Dubos had all expenses paid on his second tour of North America. He was exposed to the hospitality and attractive scenery of its greatly diverse agricultural regions. During the thirty-day trip, the delegates covered twelve thousand miles, passing through twenty-three states and four Canadian provinces. They made three dozen stops to study soil regions, experiment stations, and agricultural industries. Local newspapers along the route ran banner headlines with stories about soil experts from twenty-seven nations who came to dig pits in their indigenous soils. Russell was full of admiration for this expedition in which, he said, "men and women of widely different origin and background, who at the outset were eyeing each other with some suspicion, were before long prepared to pool their stocks of scientific and technical knowledge and help each other in the solution of their technical problems."[6]

For Dubos, this trip provided firsthand experience about how new methods and ideas in soil science needed to be tailored to each kind of native soil. Now, as one of these new microbiologists, he knew that microbes play beneficial roles in the health of the soil, but he had yet to apply this view to medical microbiology and human diseases.

After this tour, Dubos journeyed to France to see his family for the first time in three years. Aboard the *Leviathan* were many other soil scientists returning home from the congress. Two articles by Dubos appeared in the ship's newspaper, *Souvenir Log*.[7] With a light and polished touch, one called for scientific respect for "'Dirt' as Soil" and revealed the presence on board of "doctors of soil" who were carrying hundreds of "small mysterious bags" of soil samples to study in their own laboratories. Their efforts had enriched America with "knowledge of the long history of European soils," and he predicted this would lead to more fruitful science, because "it is the shock of ideas that causes sparks to fly." The second article, "The Typical Foreigner," contained a portrait of himself as the average Frenchman with "green eyes, straw colored hair, 6 feet or so, almost all of them being legs." At this conclusion to one phase of his life and ready for the next, he was in a jubilant mood.

On returning to New York and Oswald Avery's laboratory in September 1927, Dubos joined the effort to seek a cure for infectious pneumonia. He was always grateful and a little awed that such an institution "would have taken a person like me, knowing nothing at all about medicine, and coming from an agricultural experimental station, and given him a chance to work in a hospital."[8]

A Medical Gamble

Rufus Cole, the first director of the Rockefeller Hospital, had selected lobar pneumonia for study in 1909. In America, it was the leading cause of death and no effective treatment was available. Some futile remedies tried against this disease had varied from bloodletting to cold compresses, from inhalations of chloroform to injections of gold, and from doses of mercury to glasses of whiskey. An experimental pneumonia vaccine had also been tried, but since it caused adverse reactions, medical scientists continued to look for other treatments.

Cole brought Avery to Rockefeller in 1913 to establish a serum therapy program against pneumonia. Serum therapy works by providing passive immunity to patients after they become ill, in contrast with vaccines that provide active immunity by preventing disease. Patients were treated with injections of serum containing antibodies recovered from immunized horses. The patients use antibodies in the serum to counteract the antigens (the immunological opposite term of antibody) being produced in their bodies by the pneumococci. There was good evidence that this therapy could succeed. Emil von Behring won the Nobel Prize in 1901 for devising serum therapy against diphtheria. In 1905, at Rockefeller, Simon Flexner began using serum therapy against hundreds of cases of epidemic meningitis in New York City. Since serum was not commercially available, Avery was responsible for vaccinating horses with pneumococci, obtaining the antipneumococcal serum from them, and then administering it to patients.

By the time Dubos arrived, the Hospital laboratory had spent fifteen years researching serum therapy and physician-scientists had treated more than fourteen hundred pneumonia patients with very encouraging results.[9] During this time, however, Avery learned lobar pneumonia was not one disease but several, each caused by a different type of pneumococcus. Their serum treatment worked well against types I and II but not against other types, particularly the most virulent, type III. Dubos' first task was to figure out how to dissolve the cellulose coating of the type III pneumococcus. As it turned out, after he found a substance, he and Avery used it to treat type III pneumonia, first by systemic injection and then in serum therapy. His approach to this problem led to modern antibiotics.

Dubos worked full time at a small bench outside Avery's office on the sixth floor of the Hospital, unlike the M.D.s who moved between the laboratory and wards with pneumonia patients on the floors below. Spartan and utilitarian, the laboratory contained a wooden desk that, as Dubos described, "accommodated a motley assortment of notebooks and simple laboratory instruments—test-tube racks, glass Mason jars, droppers for various dyes and chemical reagents, tin cans holding pipettes and platinum loops. . . . The Bunsen burner on each desk served for aseptic transfer of cultures, heat sterilization, preparation of culture media, and also for some chemical operations. We used a great variety of kitchen utensils. . . . [There were] a few simple incubators, vacuum pumps, and centrifuges . . . [and] a single porcelain sink that served for almost any operation requiring the use of water, from staining slides for microscopic work to preparing extracts of bacterial cultures for immunological tests."[10] The serviceable nature of his laboratory did not change appreciably during his career, despite later moves to newer buildings.

Within the laboratory, Avery had a method of keeping his staff devoted to some aspect of the pneumococcus. Following Institute practice, the newcomers were neither formally trained nor assigned to specific projects. Dubos recalled, "Avery never asked anyone to do anything. In fact, he almost urged people not to do too much. Of all the persons I have known in science he certainly was the man who most was concerned with thoughts, long thoughts and meditation, before doing experiments, instead of the usual manner of rushing in and doing as much laboratory work as possible."[11]

A newcomer was drawn into endless conversations, or monologues, in which Avery spun out simple hypotheses about disease based on a few salient facts. Seasoned laboratory members dubbed these monologues "Red Seal Records" (from the famous RCA record label of classical music). After considering a variety of possible approaches to problems with Avery, a beginner was left to initiate projects that were compatible with his own technical skills and intellectual temperament. Thus each researcher became specialized in techniques for which he was best suited. "We had to discover within ourselves," Dubos said, "the initial impulse and the initiative to act as individuals."[12]

Almost at once, he was on his own as an investigator, weighing the "long thoughts" against the short list of techniques at his disposal. He learned to practice rigorous experimental science by following Avery's admonition to be "bold in formulating hypotheses but humble in the presence of facts," in other words, to have great respect for facts and exacting requirements for evidence. Due to his poor eyesight, Dubos did not trust his hand-eye coordination. "I'm not very gifted with my hands," he said, "and moreover I'm not very skilled in laboratory operations. I spend the greatest part of my time just thinking about problems rather than doing things about them." A candid summary of his strategy was that if an experimental problem is well formulated, its testing and solution should become simple and straightforward.

His experiments, in general during his career, did not require an apparatus more complex than a microscope or centrifuge and were designed to be capable of unequivocal interpretation, to the extent of eschewing statistical analysis for evaluating data, leaving no place for the observer's imagination. "Statistics," Dubos said, "are often a will-of-the-wisp which lure their eager followers into the marshy lands of indecisive experimentation."[13] His name did not appear on a paper unless he had participated in the work himself. At the bench, he was impatient for results, always eager to put experimental data into a picture that had structural quality. Unlike Avery and Louis Pasteur, he did not dramatize his research with a single "protocol experiment," an artistic performance to demonstrate an expected result without fail. His dramatizations came in lectures and essays, where he had a flair for exploring the import of scientific ideas.

Experiments before 1929 did not address directly Avery's problem with type III pneumococci. Dubos focused on a practical question that was compatible with his skills, namely, finding a way to produce large quantities of pneumococci to work with. Success came in two years and was due primarily to the ecological premise that "any problem of growth is of course a problem of reaction between an organism and the environment." Pneumococci were difficult to grow in ordinary culture media, so he manipulated environmental factors, in particular, removed oxygen so that competing bacteria withered while the pneumococci grew abundantly.[14]

Although a novice in medical bacteriology, Dubos dared relate this test tube experiment on growth to a mechanism of human disease. He speculated that tissues deprived of their normal oxygen supply would be more susceptible to bacterial invasion and thus allow infection to spread. Some evidence supporting this early hypothesis would be found nearly twenty-five years later in his laboratory studies of tissues infected with staphylococci. The ecological principle he noted here is that the prevailing biochemical environment (in tissues and in culture media) can determine whether bacteria will flourish or languish.

Life at the Institute

Within the Hospital, some young doctors coined the motto "Poverty, celibacy, and science" to portray their life of financial austerity and scientific rigor.[15]

Only partially fitting this picture, Dubos was described by a friend as "thin, shy, unimpressive, underfed, apparently underpaid in the noble Flexner tradition. You talked little but said a lot."[16] Under Cole's rules, physicians who joined Hospital laboratories and cared for patients had to live and take all their meals within the Hospital. They could not be married—thus the "celibacy" in their motto.

Although single and as poor as the M.D.s, Dubos was by training and position a Ph.D. in biochemical bacteriology and worked hard to fit into this medical conclave. He prevailed as an outsider, even a loner, while taking up a totally separate lifestyle outside the Institute. From 1927 to 1929, he lived at International House, on Riverside Drive in the Upper West Side of Manhattan. (This residence was founded a few years earlier by John D. Rockefeller, Jr., who at the time was also president of The Rockefeller Institute's Board of Trustees.) Then, as now, International House was a cultural and academic gathering place for foreign postgraduate students.

Living among nonscientists made Dubos acutely aware of his limited interests and incomplete understanding of his new language and American culture. His remedy was to make it "a point to read almost anything that came my way, almost irrespective of content," including Edna Ferber and Ernest Hemingway novels and Frederick Jackson Turner's *The Frontier in American History*. He often claimed that the "most influential book" in his life was the first one he read in America, Lewis Mumford's *Sticks and Stones*, which "taught me a new way to look at the world of reality. As a young laboratory scientist, I knew how to recognize phenomena, but chiefly within the frame of abstract theoretical concepts. As a product of French education, I had some awareness of the humanities, but in a rather abstract way." Mumford, in contrast, dealt with "subjects related to both science and the humanities—in the very earthy terms of sticks, stones, and the concrete experiences of daily life." Mumford's emphasis on quality of life and total experience was so satisfying, he added, "that I immediately knew that I had discovered a master to my taste—a guru." Many years after reading this book, when the two men eventually met, Dubos was greatly astonished to learn Mumford was only six years older than he was.[17]

Other than these bookish insights, the rare letters that survive from this period allude to acquaintances with artists, writers, and scholars far removed from the world of science. To earn extra money, Dubos gave French lessons to Columbia University anthropologist Franz Boas, charging him two dollars an hour. Unfortunately, this side of his life—*bon vivant* or bohemian?—is almost entirely lost.

During the summer, research at the Institute essentially paused. There was no air conditioning and New York City's heat and humidity were devastating to experiments. The horses used for serum therapy were sent to pastures far upstate. Staff members were encouraged to close their laboratories while writing and resting outside the steamy city.

Dubos spent four months in France during the summer of 1929. He was not well and needed the rest, telling his Rutgers friend Robert Starkey, "I really feel very tired and my heart seems to be weakening a little."[18] On the ten-day voyage, Simone Bramerel and her husband Marcel, a French chef working in New York, found the young scientist sitting outside every day on the top deck, covered with a deck robe, close to the funnels out of the wind and facing the sun. He told them his rheumatic fever, weakened heart, and arthritic pain were making him extremely weak and tired. Marcel offered to teach Dubos fencing as a good but not too strenuous exercise. So after their return to America in the autumn, the two became fencing partners at the McBurney YMCA. Marcel also prepared meals to help build his strength, even hosting a huge feast, which Dubos described as *pantagruelique*, before his marriage in 1934.

The 1929 vacation included nostalgic visits with childhood friends in the village of Auribeau on the northern slopes of the Montagne du Luberon where he walked and painted for two months. Although he was offered a scientific position in France, he told Starkey, "I am not ready to accept yet." Sailing back to New York that fall, he met Norman Taylor aboard the *Minnesota*, who was returning after a four-year stay in France. They agreed to share an apartment two blocks from Washington Square on Sullivan Street, which they fancied as the Montmartre of New York. To his American friend Starkey, Dubos enthusiastically described its view overlooking an immense private garden with lawns, trees, and a restaurant. "The furniture may be a little scant but we have wonderful paintings on the walls (mine, of course), the phone (a French one) is numbered Spring 3412, which sounds like modernistic arithmetic, and the radio allows us to hear oranges grow in California and Byrd freeze at the South Pole."[19]

In contrast with his previous trip, this return to America was painful. To his French friend, artist and writer Robert Potet, he complained of homesickness and unhappiness. "I have learned to love too much the sweetness of living that one knows only in France. The mechanical power of America has truly been a suffering for me. . . . it imposes itself violently. I have a profound antipathy for the machine and the machine triumphs here marvelous, perfect, but really too inhumane. I've taken to hating metal, I want to have around me only wood, so much more rich, savory, vibrant. The immense metallic noise of this city breaks my nerves. . . . I seek in everything, in every person, something spontaneous which is not the clear result of a premeditated design. My word, I could become a romantic again."[20]

There was also anguish about recovering enthusiasm for research, because it was bringing no intellectual satisfaction. Science, he told Potet, "is one of the only sources of adventure which the modern world can offer. I consider it more as a thrill and excitement than intellectual pursuit; . . . every culture that I put in the incubator is a test of myself against nature and each morning I verify my results, I know the emotion of the gambler who returns to his dice but I often have the impression that my activity is as futile as the one of the gambler."[21]

Such a negative statement goes beyond mere youthful fear or insecurity about whether he would solve Avery's problem. Viewing the experimental method as a gamble, and the scientist as addicted to its chances, is one of the earliest expressions of his trenchant skepticism. This may signal his initial doubts about a physicochemical approach to living processes that would grow more pessimistic. It may also express misgivings about whether science could satisfy his longing for an "intellectual pursuit," a search that eventually led him to an understanding that science becomes meaningful only after creating patterns from its profusion of facts. The notion of gambling with results further points to a dislike of "premeditated design" and the kind of control needed to follow experiments like recipes. Paradoxically, the methods that were giving a sense of mastery over nature were producing seeds of doubt about whether nature can be controlled.

Discovering Antibiotics, or Life Impedes Life

An observation at least as old as Pasteur is that germs can impede other germs. In 1877, Pasteur reported an experiment where anthrax cultures exposed to air lost their infective power, a transformation he attributed to contaminating aerobic microbes. Based on this observation, he suggested the phenomenon of *la vie empêche la vie* might one day "justify the highest hopes for therapeutics."[22]

Twelve years later, in 1889, Paul Vuillemin coined the term *antibiosis* to describe a process where, in his words, "an organism destroys the life of another organism to preserve its own." He labeled the destroyer an *antibiote* and proposed the term *antibiosis* (against life) as the converse of the term *symbiosis* (with life).[23] In the 1940s, Waksman narrowed this concept to microbes and defined an antibiotic as "a chemical substance produced by microorganisms which has the capacity to inhibit the growth of bacteria and other microorganisms and even to destroy them."[24] Today, whatever the origin of these drugs, the word *antibiotic* retains its overall meaning of living cells antagonistic to other living cells.

The path to the discovery of modern antibiotics, like most scientific advances, involved the work of many people over several decades. In the sixty years after Pasteur and before Dubos, there was significant research on bacterial antagonisms and their potential for curing disease, albeit without finding anything of practical importance. As Alexander Fleming, the discoverer of penicillin in 1929, noted in his 1945 Nobel lecture, "This fact that bacterial antagonisms were so common and well-known hindered rather than helped the initiation of the study of antibiotics as we know it today."[25] Moreover, few studies were done on bacterial properties or the biochemical nature of the active substances found in bacterial antagonisms. Waksman, a codiscoverer of streptomycin in 1944, attributed the failure to discover antibiotics any earlier to the fact that these early scientists limited their "search for substances that would kill pathogens in the test tube rather than in the treatment of systemic infections."[26]

Among the hopeful agents, some historians have made a partial claim for pyocyanase as the first true antibiotic that was produced commercially and used clinically. Introduced in Germany in 1899, this microbial product was a mixed substance produced by *Pseudomonas pyocyaneus* that had limited use as a spray to treat topical (not systemic) infections of anthrax and diphtheria. However, medical practitioners rapidly lost interest after 1914 since the agent's chemical composition was unknown; the preparations were impure (supposed enzymes), unstable, ineffective, and even toxic. Overrated claims and unconvincing scientific evidence based on such a highly variable product quickly relegated pyocyanase to a marginal remedy. As Howard Florey and other English researchers observed in their review of fifty years' work on pyocyanase, the drug "stands as a memorial to experimental and clinical inadequacies . . . both of technique and interpretation."[27]

Several isolated chemicals of nonbacterial origin had also been tried as antimicrobial therapies; most were as destructive to host cells as to the pathogens. Generally, these chemotherapies were fortuitous bits of empiricism rather than products developed through understanding pathogenic bacteria. Some bactericides of chemical origin included bile salts, coal-tar dyes, gold, silver, iodine, and cinchona, all without significant effect.[28] Others were based on highly toxic principles—arsenic, mercury, phenols, fatty acids, basic dyes. By the mid-1930s, according to Harry F. Dowling, infectious disease specialist and historian, "practically all investigators and clinicians working in the field of infectious diseases had given up hope that effective chemical agents would be found for the treatment of bacterial infections."[29]

At least two chemical agents proved successful against bacteria. Early in the twentieth century, Paul Ehrlich developed the first synthetic chemical drug, the arsenical salvarsan, for systemic use against syphilis. He also coined the words chemotherapy and charmed or magic bullets to refer to drugs that seek and destroy pathogens without harming the body. Developing his concept of a magic bullet usually started with an enterprising chemical industry taking a lethal chemical and detoxifying it by chemical group substitution or modification. A successful example of this approach came in 1935 when Gerhard Domagk discovered the sulfonamide called prontosil by following Ehrlich's idea that dyes stain living cells in much the same way that certain drugs kill them. Prontosil and other sulfa drugs were derived by synthesizing new chemicals from known dyes and then testing their antibacterial properties in animals.[30]

In contrast, the antibacterial discoveries made by Dubos during the 1930s were not simply fortunate accidents but the result of systematic searches for chemicals derived from bacteria. He took a distinct ecological approach to the study of living systems by interpreting and controlling processes of microorganisms in their environments. According to Florey, Dubos' approach "had the outstanding merit of considering the subject from many points of view—bacteriological, biochemical, biological, and eventually clinical."[31] Each discovery produced a full portrait of an antibiotic, from its isolation and purification to an analysis of how it cures disease.

The pneumococcus and the S III enzyme. The first of these antibacterial agents was the substance that solved Avery's problem. By late 1929, Dubos could produce enough pneumococci so that he could begin looking for substances capable of destroying the pneumococcal capsule, the cellulose-like coating that protected the bacteria from natural defense mechanisms of the human body. Techniques for the next step, removing, or purifying, the capsule had already been developed by Avery and sensitive and specific assays were available to determine whether the purified polysaccharide was still intact or had been destroyed. The Dubos experiments that followed are of considerable historical and practical interest in light of a renewed interest in finding antibiotics from soil microorganisms.

There is no way to know what substance Dubos expected to find. One assumption was that because type III pneumococcus was typically fatal, there was no naturally occurring enzyme in human or animal tissues that worked. This idea seemed to be confirmed in Waksman's 1926 monograph on enzymes, including microbial enzymes, that did not consider their therapeutic potential.[32] Avery believed an enzyme would dissolve the capsule and had tried every one he knew, without success; the only substance he found to work was hydrochloric acid in large amounts, a chemical that obviously could not be used in the body.

Based on the cellulose studies at Rutgers, Dubos could still assume that a selective bacterium might be found from which a substance could be extracted. He was no doubt further encouraged by animal experiments of two Rockefeller colleagues, Francisco Duran-Reynals and Jacques Bronfenbrenner. Within the Institute's small community this was the kind of science discussed at weekly staff meetings, journal clubs, afternoon tea, and particularly in the lunch room.

Both Duran-Reynals and Bronfenbrenner had recently come from the Pasteur Institute in Paris, where they worked with bacteriolytic microbes. Duran-Reynals observed that *Tyrothrix scaber*, a bacterium discovered by Emile Duclaux in the 1880s, was able to produce powerful antibacterial substances to destroy intestinal microbes. Bronfenbrenner compared *Tyrothrix* behavior on the intestinal microbe *Bacterium coli* (an early name for *Escherichia coli*) with that of bacteriophages, the viruses that prey on bacteria. He found the phages caused the coli to swell and later burst, whereas *Tyrothrix* attacked the coli's cell membrane and caused the bacteria to shrink in size and disappear. Here, then, was a relatively unknown microbe that existed in the human environment without causing disease. Further, it attacked other bacterial species and worked by breaking down the bacterial cell. While *Tyrothrix* itself did not turn up in Dubos' immediate experiments with the pneumococcal capsule, its potential was of theoretical significance and the name tyrothricin was coined for his famous 1939 antibiotic.[33]

Combining some methods in Avery's "kitchen chemistry" with those in agricultural chemistry, Dubos induced soil bacteria to digest the pneumococcal polysaccharide. He knew that bacteria perform highly specialized tasks, so the trick was to find the right species. This adaptation of Winogradsky's direct method is called, paradoxically, soil enrichment or starvation. First, various soils were chosen in which organic materials accumulate and vanish, including leaf mold, composts of corn cobs and rye straw, manured soils, and peat soils. Then microorganisms in these samples were fed one source of food, the purified capsular polysaccharide of type III pneumococcus. This is called enriching the soil. Since most microbes were unaccustomed to such diets, they remained dormant or starved. He was looking for a microbe with an appetite for the capsular material.

Success came from a soil sample taken out of a New Jersey cranberry bog. One bacterial species kept growing on this unnatural high-carbohydrate diet, and from this soil sample Dubos isolated the bacterium and then from it isolated the responsible "antibacterial" enzyme. In published reports, the bacterium was called "S III bacillus," in other words, the bacillus that attacked the specific ("S") polysaccharide of type III pneumococcus. The enzyme isolated from the bacillus was neither chemically identified nor named and was referred to in print as "the bacterial enzyme" or more informally as the S III enzyme.[34]

By spring 1930, Dubos was so pleased that he bragged to Starkey of the "rather startling piece of work which keeps us talking, dreaming, and planning. I'll tell you something about it, although it is still a secret. To excite your imagination, I may repeat some of the comments. 'It is a great discovery' (Dr. Cole). 'This experiment may itself justify the existence of the whole Institute' (Dr. Cole). 'It is a new principle' (Dr. Swift). That sounds like an advertisement for toothpaste, but it is really very exciting."[35] Starkey responded that he already knew about the experiments from Mortimer Anson, at Rockefeller's facility for animal and plant pathology in Princeton, New Jersey, who "suggested you were having a rather interesting time with the enzyme produced by the bacterium which you found was able to decompose the hemicellulose of the Pneumococcus."[36]

By early summer, the work had moved forward so quickly that Dubos wrote to Avery who was on vacation in Maine. In an atypical move, Avery returned immediately so they could repeat the experiments together. Their first announcement of this discovery was published in *Science* on 8 August 1930.[37] Another eleven months passed before the full reports were published.[38] To test the enzyme's action on the encapsulated pneumococcus, they performed more in vitro experiments before testing whether injections of the substance protected mice and rabbits from disease. They found that the enzyme by itself was neither bacteriostatic, bactericidal, nor bacteriolytic. It worked by dissolving the capsule but without impairing the viability of pneumococci or their ability to regain their capsules. Deprived of their sugar coats, the naked bacteria were quickly destroyed by the phagocytic, or digesting, cells and the animals were cured. This totally different therapy was the first time an animal had been protected against the pneumococcus except with an immunizing serum therapy.[39]

Avery and Dubos did not consider this an immunological action where capsular material would have been neutralized against phagocytosis by "union with the type-specific antibody." Instead, they regarded the enzyme as performing a preliminary chemical action to decompose the polysaccharide. "The end result, so far as the fate of the microorganism is concerned, is the same," the authors wrote, since the enzyme, like an antibody, merely initiated a protective reaction, namely, the cellular response of the host.[40] This discovery was convincing proof of Avery's hypothesis that the capsular polysaccharide was responsible for the virulence of type III pneumonia.

Dubos was sole author of the first four papers he published after joining Avery's laboratory. For the 1930 *Science* paper, however, Avery took responsibility as principal author, with Dubos relegated to a supporting role. Much to his disappointment at the time, he felt he was "just a pair of hands" and was pained when Avery received all the letters of admiration. Confiding unhappiness to Potet, in 1933, he said "I fear to not have drawn a great benefit; I made this discovery too young, the honor for it will go to my chief. I say this without too much bitterness; he is a very distinguished man and if I am really worth anything, it will fall to me to show it again much later."[41] Years later, he accepted that the "more difficult task of conceptualizing the problem" had originated with Avery years before and that his contribution was in fact the technique of carrying out the idea.[42]

Adaptive enzymes. Immediately after unearthing the S III enzyme, Dubos made an unexpected observation that he always referred to as "one of the most important biological laws" he ever encountered. This happened while developing methods to produce, purify, and concentrate the enzyme to test its effects on rabbits and monkeys. There was no difficulty in producing huge amounts of the soil bacillus when it was cultivated in a rich bouillon but—surprising and disappointing—the large microbial mass did not contain the S III enzyme. This meant growth was not linked to enzyme production. The significance—and frustration—of this experiment can be appreciated by realizing he tried more than fifty different media without success.

Eventually, he recognized that the enzyme appeared only when the bacillus was deprived of all nutrients in the media except the capsular polysaccharide or a related carbon source. "Teleologically speaking," he theorized, "the secretion of the specific enzyme appears then as an emergency measure" when bacteria are deprived of growth energy.[43] This phenomenon, he said, brought him face to face with the fact "that cells have multiple potentialities and that these potentialities usually become manifest only when the cell is placed in an environment where it is compelled to use them. . . . I take a great deal of pride in the fact that I discovered it without anyone's help or advice."[44]

These experiments again confirmed Winogradsky's observation that microbial activities were not "purely chemical effects but also biological adaptations." Dubos elicited an inborn chemical action from a specific microbial species, one that produced out of its innate machinery the key S III enzyme. What he also witnessed, in trying to get quantities of the enzyme, was an adaptive response of bacteria. Unlike "constitutive enzymes" that are always produced by bacteria, the discovery of "adaptive enzymes" (now called inducible enzymes) showed that bacteria also express a selective and reversible behavior. The techniques in biochemistry and genetics to inquire further into this phenomenon were not yet available, but they were found and used in the 1950s by Dubos' friends at the Pasteur Institute in France, Jacques Monod, François Jacob, and André Lwoff, who received the 1965 Nobel Prize for elucidating how genes direct the synthesis of enzymes.

More important to Dubos, however, was the realization that bacteria can display morphological, physiological, and even pathological properties that are normally hidden. This phenomenon provided great intellectual satisfaction because, in his words, it "brings the bacteriologist back into the main channels of biological thought, to the biological problem *par excellence*, the problem of adaptation."[45] The phenotypic expression of the bacteria was not the result of natural selection or genetic mutation but rather a biological adaptation leading to compromise and accommodation. This hard-won discovery—that organisms have multiple potentialities and make creative adaptations—became and remained the basis of his ecological thinking. The lesson revisited in his future experiments and writings taught that by varying an environment, it is possible to control which potentialities are expressed or repressed. He applied it to microbes as well as to human beings.

Turning the S III enzyme into systemic therapy. As Avery had predicted, the S III enzyme showed remarkable potential for curing type III pneumonia. The first trials using the enzyme involved testing it in mice as systemic therapy, that is, injecting it into circulating blood following intradermal infections that caused bacteremia or blood infections. The enzyme also exerted both protective and curative effects in rabbits and monkeys. However, a startling medical phenomenon emerged when the scientists tried curing lobar pneumonia instead of blood infections in these animals.

Avery did not believe mouse and rabbit experiments were adequate tests of an innovative therapy to inject enzymes, or of systemic therapy in general, so he had the physicians devise methods to reproduce a lobar pneumonia in monkeys. Dubos' role was producing the enzyme, while his colleagues, physicians Thomas Francis, Jr., and Edward Terrell, carried out experiments with the enzyme on animals and monitored the course of their disease.

They experienced two problems. For one, the scientists could detect no relation between the size of the infecting dose and the severity of the illness or recovery from pneumonia. Since they could not routinely make the monkeys equally ill, they had no way to create test and control groups. Even the most virulent type III pneumococci could not initiate infection unless the normal defense mechanisms of the monkeys were lowered. So they explored two variables, the physiological status of infected individuals and the impact of environmental conditions. Interestingly, they successfully lowered the animals' resistance by subjecting them to various environmental factors including chilling, alcohol, morphine, fatigue, and deficient diets.[46] According to Dubos, "It was well known . . . that the incidence of lobar pneumonia was in some way correlated with attendance at football games, and probably with the excessive consumption of liquor on such occasions."[47]

Another obstacle was that enzyme therapy did not work well in an individual with lowered resistance and severe lobar disease. This had been predicted by Avery and Dubos when they showed the enzyme only decapsulated pneumococci. Full recovery depended on the host's cells to dispose of the vulnerable organisms. If the host was in a severely weakened condition, this response could not be completed.[48] Nevertheless, the physicians had some success in treating type III lobar pneumonia systemically with S III enzymes. Despite its promise in animals, using the enzyme as a systemic therapy was abandoned by 1935. The difficulties included getting quantities of the enzyme and then obtaining potent nontoxic preparations. The enzyme displayed variable activity, and getting a uniformly pure preparation depended on many culturing factors.

While these experiments were not immediately useful, the results stimulated Dubos' thinking about the relationship of host and pathogen. He saw that it was not enough to bring the two together to elicit disease. Environmental factors affected individual resistance to infection, the course and severity of disease, and the efficacy of the therapy. Twenty years later, he devised a new research program to understand the difference between infection and disease (chapter 4).

At the end of his life, Dubos claimed that "the anticapsular enzyme that I prepared in 1929 was the first antibiotic produced by a rational scientific method."[49] This reflects both his mature ecological ideas as well as his understanding of bacterial antagonisms when the discovery was made. Contemporary scientists do not call enzymes antibiotics because they use Waksman's modern, technical, and limiting definition: an antibiotic must destroy or inhibit the growth of microbes. In their view, using this definition, the SIII enzyme only removes an external product of pneumococci, and other cells in the body have to ingest these naked pathogens before recovery can take place.

In the broader scientific sense that prevailed before Waksman's definition, Dubos was justified in calling the enzyme an antibiotic if for no other reason than it simply fulfilled the earlier ecological vision of Pasteur, among others, of one life impeding another life.[50] His use of the word *antibiotic* was also not a precise concept in medicinal chemistry but rather an example of what he and Waksman both understood was the "domestication of wild forms of life for the benefit of man."[51] Moreover, as Waksman later noted, "had the scientific world fully appreciated at that time the significance of this [enzyme] discovery the practical development of the antibiotics might have occurred much earlier."[52]

Laboratory Interludes

Momentous setbacks occurred after this discovery. Although the first two years at Rockefeller had been minimally productive, the gambles of 1929 and 1930 paid off. Dubos was buoyed by publication of this research during the summer of 1931 and left for vacation in France. On returning to New York, his anxiety about the future deepened. He moved to an eighteenth-floor apartment in Brooklyn with a twenty-five-kilometer view that looked down on "elevated trains, lighted boulevards, and miserable tenement houses strewn among square skyscrapers. All the avenues converge toward me, the metallic and explosive noise of all kinds of vehicles makes me long for the quiet step of a hay wagon."[53] The noise irritated him and the rheumatic pains persisted. While he had received two reappointments in Avery's laboratory, moving from fellow to assistant in 1928 and to associate in 1930, he worried about how the Great Depression would affect his reappointment and his salary—since the Institute had suffered significant losses in its endowment. There seemed to be no projects to justify staying at Rockefeller.

A bleak period ensued. From March 1932 to August 1935, Dubos published nothing. The two papers on which his name appeared during this time were reports of animal tests done by the M.D.s, for which he only produced the enzyme. Lapses in research naturally occur, but this hiatus was unusual.

This period involved several aspects of the baffling rheumatic fever that affected Dubos both scientifically and personally. At the time, the disease was a forbidden topic, a kind of social quarantine that kept its victims from leading a normal life. A popular perception of the disease was illustrated at the 1939 New York World's Fair in an exhibit devoted to rheumatic fever (prepared by Homer Swift of the Institute). It was headed "Cripples Who Do Not Limp" and deemed the disease a terminal illness that could cut short a life or career at any moment.[54]

Well aware of limitations for employment or promotions for someone with this dread disease, Dubos shielded any signs of physical weakness or disability. Outwardly he tried to portray a certain mirage of being healthy. Details of his precarious situation were known to few, yet some events during such a difficult and humbling time show a lot about his intense, resilient nature. Out of this personal experience eventually came his confident advice to others that "solving problems of disease is not the same thing as creating health and happiness."[55]

The enigmatic streptococcus. The idea that a streptococcus might be the cause of rheumatic fever emerged in the early 1900s.[56] For years, this etiology could not be tested because no one knew how many kinds of streptococci existed, which ones were dangerous to humans, or how they damaged tissues. Investigators were able to recover streptococci from patients with severe tonsillitis but they could not find any one type in patients with rheumatic conditions that followed several weeks later. In laboratory experiments, moreover, they could not, and to this day cannot, reproduce the disease in animals.

During World War I, epidemics of streptococcal (different from pneumococcal) pneumonia killed many military personnel following infections of measles or viral influenza. In 1918, Avery and Rockefeller colleague Alphonse Dochez returned from a commission ordered by the Surgeon General with a collection of streptococcal samples isolated from patients in Texas military camps.[57] They hired Rebecca Lancefield, a graduate student in bacteriology, to determine whether the streptococcus, like the pneumococcus, had distinct types. Within a year, four types of hemolytic, or pathogenic, streptococci were identified.[58]

Lancefield's success led the Institute to establish a laboratory in 1919 devoted to rheumatic fever. It was headed by Homer Swift, whose reputation rested on making the first American studies of treating syphilis with salvarsan. He believed that rheumatic fever was not caused by hemolytic streptococci but by the allergy or hypersensitivity of the tissues to streptococci or their toxins.[59] This new laboratory shared the sixth floor of the Hospital with Avery, so Dubos had an opportunity to see on a daily basis the obstacles that scientists confronted while trying to find the cause of his own rheumatic fever. Swift also became Dubos' physician.

Perhaps due to these encounters, Dubos turned from his studies on the S III enzyme of pneumococci to studies where he tried "to isolate new enzymes for other types of pathogenic microbes," very likely one to attack the hemolytic streptococci.[60] This task seemed more pressing when Alvin Coburn, a resident physician at Columbia Presbyterian Hospital, discovered in 1931 that a hemolytic streptococcus was indeed the cause of rheumatic fever.[61] According to Swift's annual report for 1931, Lancefield was trying to decompose virulent hemolytic streptococci that appeared to have a carbohydrate capsule similar to the pneumococci. Her goal was similar to what Dubos was doing to the pneumococci, so his involvement in her project seemed quite natural, even a self-centered gesture. Since the surface structure of streptococci was so poorly understood, and no purified forms of its capsular material existed, the experiments had to begin by using the dangerous, intact bacteria.

Dubos was energized and puzzled by the new research, telling Starkey in June 1932 that his work "has been much slower than expected and I am trying to push it ahead if possible."[62] He decided not to go abroad that summer to keep working on the problem. "I have been delaying doing anything from week to week in the hope of getting a few things settled. However, I am beginning to feel that, as once stated by Dr. Carrel, no good work has ever been done between June and October."[63]

Quite suddenly in the spring of 1933, Dubos fell victim to his fourth and final bout of rheumatic fever. It would not be idle speculation to suggest this episode may have been caused by his close contact with the pathogenic streptococci.[64] Depressed, perhaps even embarrassed, he endured this fourth attack on the very frontiers of scientific medicine. This time his caring physician Swift taught him something about the art and science of medicine.

Under Swift's care, Dubos was a patient in the Rockefeller Hospital from March until July 1933. To Starkey, he described it as "an exceedingly severe streptococcus tonsillitis and infection of the sinuses, following which developed rheumatic muscular pains and some heart deficiency."[65] The only treatments were still prolonged bed rest and kindly admonitions to relax. He was not permitted to leave his room, not even to visit his colleagues in the laboratory. During this hospitalization Swift removed Dubos' tonsils, which brought a grateful end to rheumatic joint pains for the rest of his life.

Swift was such a deliberate, painstaking physician-researcher that he earned the nickname "Speedy" from his colleagues. He kept patients in the hospital as long as the illness continued and recorded their every diagnostic and pathological sign. He was famous for creating gigantic charts to tabulate, graph, and compare data among individuals. Some charts spanned months or even years and showed movements from acute disease to recovery then back to relapse and another recovery. An example of these charts accompanied his 1939 exhibit at the World's Fair. Joined by Rockefeller cardiologist Alfred Cohn, he used the newly introduced electrocardiograph in rheumatic fever patients. This sensitive instrument routinely showed heart damage to be more common and extensive than physicians had previously realized.[66]

Reflecting the Institute's new efforts in clinical investigation, Swift was also teaching that medicine was not scientific enough. It is, he said, "the harmony of nature that we wish to understand. Harmony postulates relationships." He advocated combining scientific and artistic analyses to treat all the symptoms of disease. "Not all the art of medicine is at the bedside," he said, "nor all the science of medicine in the laboratory."[67] Swift's optimistic philosophy considered the specific needs of the patient as an individual. He recognized it was not enough to protect the body from harmful factors or agents. It was also necessary for the physician to help a patient achieve a certain harmony by adjusting to changes taking place within his body. To Dubos,

Swift's warnings to relax meant he had to learn how to integrate chronic heart disease into a normal life and to adopt a more prudent lifestyle. Living with disease could be managed. More important, he learned health requires never-ending efforts to adapt to an ever-changing environment. On leaving the hospital, he sailed to France and rested for a few more months.

Before antibiotics were developed, Dubos continued to fear infections of his damaged heart. Studies done by other scientists in the late 1940s showed that penicillin eliminates the initial strep throat infection and in turn prevents rheumatic disease.[68] The antibiotic also cures infections of damaged heart valves that previously had fatal outcomes, and in fact twice saved Dubos' life. The general understanding about rheumatic fever today combines Coburn's view that hemolytic streptococci cause the acute attacks with Swift's view that allergy leads to lasting heart damage. However, no cure yet exists for the disease once the rheumatic symptoms appear. In 2000, a century after research began on rheumatic fever, Maclyn McCarty, who became head of the rheumatic fever service on Swift's retirement in 1946, concluded, "We don't know very much at all about this infectious disease."[69]

An additional complication in the laboratory during 1933 was Avery's depression and irritation, which was traced to an overactive thyroid or Graves' disease and led to a thyroidectomy followed by several months' recuperation at the residence of Dr. Harry Bray, superintendent of the Ray Brook sanatorium in the Adirondacks. Little research was accomplished in the laboratory during Avery's absence.[70] No annual report was written in 1934, and in his report the following year Avery did not include work by Dubos, whose fallow period persisted through 1935.

Finding health and happiness. Despite fragile health and discouragement, a new life opened. Dubos married Marie Louise Bonnet in 1934. The similarities of these two young lives were uncanny. Both were French immigrants who pursued graduate studies in America. They were attracted to art museums and music concerts in New York. They took long walks in the Hudson River valley to become acquainted with American trees and plants. In their love of travel, they made a grand tour of France (1935) and two cross continental automobile trips (1936 and 1938) to explore the *hauts lieux* and monuments in French and American landscapes.[71]

Tragically, they both also suffered from rheumatic heart disease, although Marie Louise was more severely affected. Their relationship, prematurely ended by her death in 1942, seemed to radiate a special, even symbiotic, *joie de vivre.* The meager details that exist for these few years confirm how important she was for bringing his life into a new focus and helping him find a different kind of health after his fourth bout of rheumatic fever.

Marie Louise, known as Malou, was born in 1898 in Angoulême and raised near Limoges. At an early age she contracted tuberculosis, reportedly from her father, who was a painter of china and died young from the disease. Although this childhood disease healed, she suffered many attacks of rheumatic fever that resulted in a severely damaged heart, difficult breathing, and pronounced restlessness. When she came to the United States in 1920, she was a proficient pianist and taught French at several women's schools, including Smith College, Hollins College, and (from 1926 to 1938) The Masters School in Dobbs Ferry, New York.

When the couple met in late 1932, Marie Louise was just completing her master's thesis in French on symbolist poets and embarking on a doctoral degree at Columbia University. She lived at nearby International House, where René had lived from 1927 to 1929, and where they probably met.

While reminding his mother that he had never professed an enthusiasm for marriage, possibly due to his disease, Dubos introduced Marie Louise in a letter as such an agreeable companion that he could count on a happy life. He added she would bring to their family "a great element of charm, sweetness, imagination and creativity."[72] The intimate wedding, attended only by four people, took place on 23 March 1934 in the Church of Notre Dame, near the Columbia campus. This ceremony was remarkable for its benediction on the couple, he wrote his mother, because the final blessing "at my request, took place in the grotto behind the central altar of the church of Notre Dame built against a rocky cliff."[73] What he did not tell her was that this rocky grotto is a replica of the sacred shrine at Lourdes, France, renowned for its powers of healing the sick.

The Duboses, according to their friends the Bramerels, were "a very loving couple." They pursued separate careers but spent all their free time together. Their tranquil life in Dobbs Ferry, twenty miles north of Manhattan, was enhanced by views of the Hudson River Palisades from their windows. Here at last Dubos found "moments of genuine happiness, far from noise, from the world, before this immense landscape."[74] As he lamented in a letter to Starkey, "I wish more clearly that circumstances should permit me to live in the peace of the country and to forget forever crowded streets, subways and elevators. But I am afraid—convinced—that such will never be my lot and I am not as yet wise enough to reconcile myself to it."[75] For the moment, and despite the lament, a certain contentment allowed him to renew his health and get back to productive science.

Adaptive Enzymes as Tools

During Avery's absence and Dubos' hiatus, the laboratory drifted away from a focus on the type III pneumococcus. A typical reaction from Avery when experiments failed or unexpected results occurred was to end discouragement with the admonition, "Now boys, whenever you fall, pick up something."[76] This is what Dubos did as he picked up two projects from his languishing work. He wrote Starkey on 26 December 1936, "I am about to break my scientific silence of 4 years with a series of several papers on the effect of different enzymes—of bacterial and animal origin—upon the morphological and antigenic structure of pneumococci."[77]

Turning the S III enzyme into serum therapy. With these projects, Dubos returned to the type III pneumonia problem, this time by trying to make a vaccine for serum therapy using the S III enzyme he discovered. A major obstacle in creating a vaccine was the autolytic or suicidal system of the bacteria that destroyed their ability to elicit antibodies. Dubos took a pneumococcal enzyme, identified by Avery's laboratory in 1925, and after purifying it was able to establish its role in the autolytic process and in eliciting broad immunity (its nature remained unknown).[78] He then devised a vaccine by exposing the pneumococci to heat and formaldehyde, which prevented their autolysis and preserved their antigenicity. In Avery's hands, serum therapy against types I and II pneumonia was reducing mortality from an untreated average near 30 percent to under 8 percent of treated patients. By 1938 a stable vaccine for producing serum in rabbits against type III pneumonia had been developed and tested successfully in Hospital patients, reducing mortality from near certain death to about 30 percent.[79]

By this time, physicians around the country were beginning to adopt serum therapy to treat pneumonia.[80] This treatment was unwieldy because it was necessary to type the patient's pneumococcus strain beforehand, there were some adverse reactions to the serum, and it was expensive. Serum therapy was also being instituted by the pneumonia service of the New York City Health Department, which operated a twenty-four-hour typing and supply service. However, large-scale production of serum therapy and its use in clinical medicine were eclipsed when the sulfa drugs showed promise for treating bacterial infections. As soon as these new drugs became available, Avery assigned physicians to compare them with serum therapy and to test combining the two treatments.[81] Serum therapy against pneumonia halted completely once penicillin became widely available in the 1940s.

Ribonuclease. Even with the best vaccines, it was difficult to formulate sera with a high level of antibodies against type III pneumococci. To study this problem, Dubos followed an early speculation by Avery that some animals naturally produce antibodies to neutralize pathogens.[82] He and R.H.S.

Thompson isolated an enzyme, found in many animal tissues, that changed the staining property of a heat-killed pneumococci from gram positive to gram negative and ended their ability to stimulate antibody production if used in a vaccine. Further investigations led them to discover that this enzyme attacked one of the body's basic building blocks, ribonucleic acid, or yeast nucleic acid as it was then known. They suggested this enzyme may have been the substance found by Walter Jones in 1912 that digested pancreas cells.[83] Dubos named the enzyme *ribonuclease* and concluded nothing could be done to counter its effects in vaccine production.[84]

While Dubos did not go on to improve the vaccines, this discovery illustrates fascinating aspects about basic research. For one, it pointed to a weakness in serum therapy. The antigens in the pneumococcal capsules were complex structures: one part is common to all types while another part is unique to each individual type and confers its immunologic specificity.[85] Immunizing with type-specific vaccines (as in pneumococcal serum therapy) was in reality impractical for it involved giving every patient a tailor-made drug. The need instead was for vaccines that transcend specificities and act against all strains of a bacterial species, a goal that has still not been met. Pneumococcal vaccines in use today are conjugate vaccines, which combine antigens of only some of the most invasive and harmful types.

In addition, ribonuclease, a small enzyme containing 124 amino acids, became a workhorse of molecular biology and justly famous as "the Rockefeller enzyme." After its discovery by Dubos, Moses Kunitz crystallized ribonuclease and Alexandre Rothen determined its molecular weight, both in 1940. Maclyn McCarty used both ribonuclease and Dubos' S III enzyme during the purification of DNA to help Avery prove it is the genetic material. Several years later, Stanford Moore and William Stein mapped the chemical structure of ribonuclease, and Bruce Merrifield selected it as the first protein he produced using his solid-phase method of protein synthesis, discoveries for which these three scientists won Nobel Prizes in chemistry.

Enzymes with clinical applications. Dubos, along with many scientists, realized that since enzymes perform such limited, well-defined tasks, they could be used as reagents to analyze many biological and chemical problems. With Benjamin Miller, he isolated one enzyme that decomposes creatinine, a normal product of muscle metabolism, and then another that converts creatine into creatinine, which is necessary for metabolism and energy production. In an unusual step, the two men turned theoretical science into a clinical test that physicians still use to measure renal efficiency by determining creatinine levels in blood.[86]

The successes in discovering enzymes that were useful tools came to the attention of the Institute's two new young leaders. In 1935, Herbert Gasser took over from retiring Simon Flexner as scientific director, and in 1937 Thomas Rivers replaced retiring Rufus Cole as physician-in-chief. They began to modify the organization by giving more independence to a few young scientists. Gasser wanted Dubos' work with enzymes "pushed more intensively." By 1938 the work merited a promotion from associate to one of the six associate members, his first advance in eight years. Dubos again found himself in an awkward position because associate members were rarely promoted to full member. Gasser considered the position "temporary and probationary" but one where an investigator showed "enthusiasm, insight, initiative, and the ability to go ahead on his own account." The Institute centered its organization around members, while associate members were nurtured to move on and take leading roles in medical research at other institutions.[87]

By this time, Dubos was becoming known in the Institute. While constantly struggling to overcome shyness in interpersonal encounters, he was never shy about interrupting with ideas and seems to have acquired Avery's talent for discussing scientific problems with everyone who would listen. At the regular Friday staff meetings, where a member presented an account of his experimentation, immunologist Merrill Chase remembered Dubos would not let go of the question of why pneumococci lose their gram-positive stain when they are crushed. Chase hastily judged that Dubos "was playing to Drs. Cole and Flexner" but soon learned this query was justified because the loss of staining "remained a nagging problem: nothing was known of cell-wall chemistry."[88]

Rollin D. Hotchkiss joined the Avery laboratory in 1935 immediately after completing his Ph.D. in biochemistry at Yale University. To him, this was a period of "magic moments when ideas are brewing." Dubos, just ten years his senior, seemed like an established investigator who knew about everything and spoke out on things about which he felt strongly. "In those years he relished every new idea . . . his open-mouthed smile was infectious enough; his wistful wondering attitude was even more so. . . . Conversations were good, two-way, and could be continued at lunch. For a number of early years, he and Alfred Mirsky were constant lunch companions. Both knew a lot about a cultural, literary, and broadly scientific world."[89]

While gaining respect as a scientist, Dubos also gained a reputation of *provocateur*. His critiques could cause novice speakers much anguish at the Monday evening Hospital Journal Club. It was, Hotchkiss said, "character building to see one's construction tumbled into a pile of bricks and then expertly reassembled by René into an architectonic masterpiece . . . in the end, it sharpened one's faculties." Rockefeller colleagues provided an ideal assembly to try out new ideas and to spark others into scientific debates.

Despite a renewed fervor for science, his private life became difficult when Marie Louise became a patient of Homer Swift. Her continually decreasing heart function and increased agitation from rheumatic complications put her in the Rockefeller Hospital for two months in the spring of 1937. After their quiet and comparatively short summer vacation in New England, she resumed teaching in the autumn, even though she became easily fatigued. The following spring both Marie Louise and René were enormously stressed when her sister Jeanne, who was living with them at the time, became incurably insane; arrangements for her care in Belgium were made by Marie Louise's sister, a Benedictine nun, Sister Solange. The Duboses' anxieties continued to increase amid spreading international conflicts that threatened war and the safety of their families in Europe.

In 1938, Dubos organized a symposium on bacterial enzymes for the annual meeting of the Society of American Bacteriologists in San Francisco (now the American Society for Microbiology). He and Marie Louise took advantage of the occasion to drive their old Ford, in the company of their Corsican friend André Draghi, through the American West. Photographs taken of Marie Louise at Old Faithful in Yellowstone Park and at Buffalo Bill's monument on Lookout Mountain show her as lively and laughing. During the journey, however, she began to cough up blood and several weeks after their return was diagnosed with tuberculosis. At the end of this troubled year, on 5 December 1938, Dubos was naturalized as an American citizen.

Launching the Antibiotic Era

> *The isolation of gramicidin and tyrocidine was the stimulus which flooded with bright light the whole previously unillumined field, and which was to set in motion the whole marvelous plan of study and application of antibiotics.*
>
> SELMAN A. WAKSMAN, 1951

> *Tyrothricin . . . was thus the beginning of the "antibiotic age."*
>
> HENRY WELCH, Director, Antibiotics Division, FDA, 1953

Avery became intensely depressed by the emergence of sulfa drugs and, according to Dubos, "felt he'd wasted his entire life" on serum therapy.[90] In contrast, the arrival of sulfa drugs probably motivated, even challenged, Dubos to "pick up something" else. Whether the catalyst in 1937 was to compete with sulfa drugs or to find a cure for Marie Louise, he dipped into the soil to find something else to attack pathogenic bacteria. Once again taking advantage of bacterial antagonisms, he followed what Hotchkiss called "the supremely simple hypothesis that soil as a self-purifying environment could supply an agent to destroy disease-causing bacteria. . . . The stage was set for kinder, gentler drugs and at first it seemed that Dubos' agent could be one."[91] Two years later, his discovery of tyrothricin, gramicidin, and tyrocidine launched the antibiotic era.[92]

The new experiments followed the same philosophy and techniques that led to finding the polysaccharide-splitting S III enzyme. This time, instead of feeding soil samples a single component like the capsule, Dubos fed the soil whole or intact bacteria that cause various human diseases. He used gram-positive bacteria, assuming their affinity to absorb the same staining dyes meant they had the same structure. The reasoning—wrong as it turned out—was that if microbes attacked these bacteria they would first decompose the cell walls, an idea previously suggested by Bronfenbrenner's experience with *Tyrothrix* in 1928.

The tumblers of soil samples were dampened with a nutrient soup composed of staphylococci, pneumococci, and hemolytic streptococci. After months of patience, optimism, and many tumblers of soil, new microbes appeared that were devouring these pathogens. Removing a sample of the soil, Dubos could place it on a glass slide and watch under the microscope as a tiny rodlike organism attached to the pathogens and caused them to shrivel and disappear. Contrary to expectations, the good microbes did not digest the bad ones, but instead secreted a substance, killing the pathogens by setting their autolytic enzymes into action and causing them to disintegrate. Next, he isolated the bacillus that attacked these pathogens and from it extracted an active antibiotic material. The first announcement of this discovery appeared in February 1939. Three detailed reports in July described how to prepare the material and how it worked to cure mice.[93]

Dubos turned the chemical analyses of the material over to Hotchkiss. While growing gallons of *Bacillus brevis* and manipulating liters of hot organic solvents, he reduced the material to a crude brown "sticky mass as unpleasant as so much uncouth earwax. But it was powerful wax all right," although not useful until refined by fractionation, first into tyrothricin, then into tyrocidine and gramicidin. Each step of fractionation was followed by "test tube and mouse assays to see where matters stood." Hotchkiss described their collaboration in *Launching the Antibiotic Era*: "Arriving at the lab each day a good hour before I did, René would absorb the information the night had brought. From my end, I could usually tell him whether the mice had looked feverish or nonchalant—noncalorific—at midnight. With all night to think about it, my cautious evaluations had usually caught up to his instant insights of the next day, so we quickly made our conclusions." Amazingly, within a few months, Hotchkiss was separating two agents and soon enough had purified colorless crystals of both, "first, the roughneck lysin [tyrocidine] . . . and then, on November 4, 1939, the gentle protector, which we named gramicidin."[94]

Names given to the early substances still cause confusion. Gramicidin—literally, killer of gram-positive organisms—was initially applied to the first crude extract, but when two crystalline polypeptides were found, they were called gramicidin and tyrocidine.[95] Tyrothricin was then chosen as the name for the extract containing both gramicidin (10–20 percent) and tyrocidine (40–60 percent).[96] Tyrothricin is based on the Greek words *tyro* for cheese and *thrix* for thread. The name acknowledges the bacterium *Tyrothrix* that had been associated with cheese by Duclaux in France and with intestinal digestion by Bronfenbrenner and Duran-Reynals at Rockefeller.

These names are more remarkable for reflecting Dubos' beneficent ecological attitude toward microbes and his medical philosophy that the soil was a healing source; he had simply "domesticated" its microbes to control human disease. From the beginning, he never used military metaphors that have become clichés in discussing disease—drugs as weapons, pathogens as enemies, or patients as fighting wars. He believed his substances were beneficial to life, counterparts to those microorganisms that ferment foods such as wine, bread, and cheese or those used in vaccines to bolster immune systems and to ward off infections. These antibacterial agents were meant to control an individual's disease but not to eradicate populations of microbes from the earth.

Another telling point is that Dubos and Hotchkiss knew enough about their substances to reject the negative words *antibiosis* and *antibiotic*, meaning "against life." To them, these terms did not account for the selective ways in which these substances attacked pathogens. They assumed other agents yet to be found would have different mechanisms of action, affect different bacterial structures or processes, and produce different levels of antisepsis. Instead of magic bullets, they believed tyrothricin and gramicidin produced more than one effect; after striking one part of a cell they triggered a cascade of effects in multiple parts of the cell, tissue, or organ. Therefore, the two scientists used the more imprecise words *antimicrobial* or *antibacterial* to refer to their agents.[97] Hotchkiss liked to recount a paradoxical feature of this discovery, namely, that deriving medicines from the soil went completely counter to a warning that said to avoid disease it was necessary to avoid soil and other unclean or dirty matter. The soil in this case could heal.

Surprisingly, the antibacterial substance isolated from *B. brevis* was not an enzyme with a specific action that disintegrated the cell membrane, as Dubos had expected. Instead it was a polypeptide with a wide range of activity. A great deal of effort went into showing how the drugs worked. Hotchkiss determined the two polypeptide fractions are not closely related in chemical structure and differ in their biological activities.

The useless, impractical tyrocidine reacts with many proteins, can be inhibited by many tissue components, and is toxic to a variety of living cells. A striking thing about tyrocidine is that its action in the test tube is against gram-negative bacteria, while its action in the body is only against gram-positive bacteria. The valuable gramicidin, on the other hand, acts on gram-positive bacteria, is inert toward proteins and many tissue cells, inhibits one or more metabolic reactions, is poorly soluble in water (thus limiting its method of administration), and is inhibited by products released by gram-negative bacilli. Tyrocidine acts as a bactericide, or general killing agent, whereas gramicidin acts as a bacteriostat that inhibits growth without killing.

There were great successes in curing infected animals with gramicidin. In mice infected with pneumococci, it was a hundred thousand times more potent than the sulfa compounds. It also cured bovine mastitis, a streptococcal disease. Elsie, the famous Borden cow, stricken with mastitis at the 1939 New York World's Fair, was one of the first "patients" to respond to the antibiotic. Veterinarian Ralph B. Little and Dubos showed gramicidin was of considerable importance to farmers, because it eradicated infection without damaging tissues or milk yield, and it eliminated the need to remove diseased animals from the herd.[98] Whether these antibiotics were tested against *Mycobacterium tuberculosis* is unknown.[99]

Within months, everyone on the sixth floor of the Hospital was immersed in the new discovery. In Swift's laboratory, Rebecca Lancefield performed animal tests with the antibiotic against ten strains of hemolytic streptococci. In all cases, she found it gave the same order of protection against streptococci as Dubos obtained against pneumococci.[100] George Hirst discovered a bacterial enzyme that removed the streptococcal capsule in vivo, following Dubos' earlier abandoned search, but unfortunately this enzyme would not cure hemolytic streptococcal infections.[101] Swift also tested the new extract against meningococcal and gonococcal infections but got less impressive results.

In Avery's laboratory, animal tests were carried out by physicians Colin MacLeod, George Mirick, and Edward Curnen. They were the first to detect dangerous side effects of tyrothricin. Although mice were protected against infections, in similar tests performed on larger animals, dogs in this case, tyrothricin not only did not work but destroyed red blood cells.[102] Avery most likely had anticipated testing the drugs in Hospital patients, because he hired Alvin Coburn, the physician who discovered the streptococcal cause of rheumatic fever, but the discouraging results in dogs immediately precluded this project.

As it turned out, Coburn, Hotchkiss, and Dubos spent two years trying to learn why tyrothricin did not work against gram-negative bacteria. The answer added a deeper level of microbial ecology to their understanding of infection. After chemist Jordi Folch-Pi, in Donald van Slyke's Hospital laboratory, pointed out that cephalin, a fatty compound in tissues and blood, suppressed gramicidin, Dubos isolated an endotoxin from gram-negative bacteria with similar qualities. When he added cephalin or the endotoxin to cells treated with gramicidin, the antibiotic stopped working and the normally susceptible cells kept growing. This experiment was an early warning to Dubos about how other biological materials compete in infections in the body and could contribute to bacterial resistance to antibiotics, among other side effects. As a further reminder, infectious processes in tissues, like soil microbes in natural habitats, were beginning to look more intricate than those studied in artificial culture media.[103]

Gasser believed the Dubos discovery had rejuvenated the Hospital's mission. In a special report to Rockefeller's board in October 1940, he said that "Drs. Dubos and Hotchkiss have kept the Hospital in the front line of chemotherapeutic attack against infectious diseases by the discovery, purification, and crystallization of a new chemical substance capable of attacking all gram-positive organisms so far tested." He further claimed their enthusiasm was not dampened by the negative findings because "they realize that they possess a new lead for the chemotherapeutic treatment of infectious diseases in general."[104]

Such a vast interest developed in the new antibacterial agents, despite their adverse effect on red blood cells, that within months almost every American pharmaceutical company began producing and testing them. A policy of the Institute, in keeping with the philanthropic purposes of the institution, was to make all its discoveries "freely available to the public." In cases involving drugs with therapeutic value, beginning with tryparsamide against sleeping sickness in 1919, the Institute filed a few patents to insure the purity of the products and to avoid their exploitation. The Institute neither participated in their commercial preparation and sale nor received royalties or other benefits from the drugs.

Hotchkiss recalled the 1940 patent application on the new agents as a hilarious series of negotiations between the unworldly scientists and the worldly lawyers, in which the contents of their notebooks were translated into thirty-six patent claims. When they were called to the business manager's office to assign the patent to the Institute, they read the clause "acknowledging the receipt of the sum of one dollar." Curious, they asked whether they meant the part about the dollar, and "The business manager, Edric Smith, mumbled something embarrassedly and withdrew, returning a bit later . . . carefully shepherding into the room two half dollars, one for each of us! Dubos, who knew intuitively how to make a grand gesture when the time was right, was positively delighted with the evident embarrassment of all concerned."[105]

Clinical research. Three foremost physicians in the United States, Charles Rammelkamp and Chester Keefer in Boston and Wallace Herrell at the Mayo Clinic, reported the first clinical uses of the local application of gramicidin at the 5 May 1941 meeting of the American Society for Clinical Investigation in Atlantic City.[106] The agent was so novel that their priorities were establishing dosages, routes of administration, and the kinds of infections that would respond to these drugs. The published reports left no doubt that they were effective in a variety of infections.[107]

These physician-scientists also quickly confirmed the darker side of the picture during their research. They found toxicity to red blood cells when the drugs were injected intravenously into mice. Remarkably, no toxicity was found when the substance was administered intraperitoneally, subcutaneously, topically, or orally. This meant the drugs were safe and effective in the body if they did not enter the blood stream. In fact, Rammelkamp found the antibiotic was especially effective and safe for local infections such as open wounds, ulcers, boils, carbuncles, and in body cavities such as the lungs and sinuses. The drugs were either not adsorbed from these sites or they were neutralized by the local tissue environment. During World War II tyrothricin was used successfully as a topical application for hundreds who were wounded in the Pacific arena. At the time, it was stable, less expensive, and more readily available than penicillin.[108]

The clinical researchers also found patients who developed resistance to tyrothricin. In an extreme case, Rammelkamp observed resistance during the fifty-third week in a patient undergoing intensive therapy for a leg ulcer; after ten more weeks on sulfanilamide therapy, the staphylococcus resisted both the sulfa drug and tyrothricin.[109] Early reports from other scientists, including Hotchkiss, suggest that resistance was not that rare and was even to be expected considering the well-known variability of bacteria.[110]

These clinical trials alerted Dubos to major aspects of infections that he had not anticipated. One was how profoundly infections in isolated, local tissue environments differ from those in circulating blood. Another was how the antibacterial agents introduce a new complexity to infections, acting in ways that could not be predicted from simple host-pathogen interactions. He immediately grasped that microbes are as adaptable (and resistant) in human bodies as they are in the soil.

Today, gramicidin and tyrothricin remain in the pharmacopeia of many countries outside the United States. They are used in ointments, ear and eye drops, and throat lozenges. Gramicidin is also being used as a research tool. Since it works by puncturing holes in bacteria, it helps scientists study how substances channel through bacterial membranes. Current research may also lead to making these original antibiotics more selective in what they destroy and perhaps expand their therapeutic usefulness.[111]

Following the Momentous Discovery

The initial publicity about the new antibacterial agents was modest and factual. The Rockefeller Institute did not issue publicity about this discovery, leaving its promotion to come from peer reviews in the scientific community.[112] In Dubos' case, the May 1939 issue of the *Journal of the American Medical Association* summarized his experiments for physicians without giving scientific details. The article added the hope that soil bacteria could be found to treat tuberculosis. The news was reported by *The New York Times* in July, the month that Dubos' full scientific reports were published; this story called the new chemical more powerful against streptococcal infections than sulfa drugs.

Press coverage escalated after Dubos spoke at the Third International Congress for Microbiology at the Waldorf-Astoria in New York. In September 1939 nearly sixteen hundred delegates representing forty-six countries gathered just as World War II broke out in Europe. Its president was Thomas Rivers, Rockefeller's physician-in-chief, who was also a prominent member of New York City's Board of Health. Fiorello H. La Guardia, New York's mayor, opened the meeting with a long speech, praising science for its role in preventive medicine and public health. He ended with an anguished hope that one day war would be considered "as odious as the nasty little microbes you discover from time to time."[113]

The Rockefeller Institute hosted a gala reception as well as laboratory demonstrations for visiting delegates. In a manner reminiscent of Avery, Dubos brandished a bottle of "a whitish gray powder" of the new antibacterial substance in front of the microbiologists and journalists in his laboratory. He boasted it was enough to protect five trillion mice against pneumonia and streptococcal infections. Such showmanship, typical of his boldness when he was enthused, was a bit rash since the antibiotics were still unnamed, tested only in mice, and of unknown composition. Within the week, *The New York Times*, among other large city papers, featured his discovery in two articles. One announced the opening of a new field of research to fight "bacterial enemies" and the other linked the amazing success to Dubos' previous experience with soil bacteria.[114]

Dubos' influence on penicillin development. Alexander Fleming, the Scottish bacteriologist who reported the discovery of penicillin in 1929, also spoke at the September 1939 Congress, but not about penicillin.

Fleming's role in introducing penicillin into medicine has been amply documented.[115] In brief, the penicillin story falls into two parts, separated in time by nearly a decade and carried out by two different sets of investigators in different institutions. The first phase, at St. Mary's Hospital in London, involved Fleming, who observed the blue-green penicillin mold on the agar plate. Because of his 1922 discovery of lysozyme, an enzyme produced naturally by tears and other body fluids that dissolves certain bacteria, he had methods for testing the inhibitory action of bacteria and evaluating their toxicity in tissues. Following his initial report, Fleming and physician C.G. Paine in Sheffield tested penicillin as a topical antiseptic against human infections but they did not publish their limited anecdotal successes.[116] After taking a few steps to isolate penicillin from the mold, Fleming abandoned the work around 1932 when the crude filtrates could neither be purified nor stabilized. However, he recommended using penicillin "as a bacteriological weed killer" in culture medium; this was a clever way to isolate certain bacteria while killing off all other bacteria susceptible to the drug.

By the time sulfa drugs appeared in 1935, Fleming had turned to other research. His topic at the 1939 Congress was the synergistic action of sulfa drugs and serum therapy on pneumococcal infections. As related by Dubos, Fleming told him at the Congress that he saw no future in his unstable penicillin, in Dubos' new agent, or in any other naturally occurring antibacterial substance. Fleming believed instead that future therapies would be synthetic chemical successors to the sulfa drugs.[117]

The second phase of the penicillin work was carried out under pathologist Howard Florey at the Sir William Dunn School of Pathology at Oxford University. Its development into an antibiotic was greatly encouraged and enhanced by Dubos' work, which had produced a result that Fleming had failed to achieve. In contrast to what Dubos remembered in later years, Florey did not attend the 1939 Congress.[118] However, news of the Dubos discovery had already reached him in England and was instrumental in his and Ernst

Chain's decision to study antibacterial substances of microbial origin, in particular to revive Fleming's dormant research on penicillin. Florey was in Oxford that September, where he immersed himself in correspondence with the Medical Research Council. He was trying to raise funds for studies on "substances with chemical properties very similar to those of lysozyme which act powerfully on pathogenic bacteria." Florey cited the Dubos agent in his 6 September proposal as a good precedent for investigating the field of antibiosis: "[O]nly the bactericidal principle from soil bacteria which is especially effective against pneumococci has been studied in some detail. Recently prominence has been given to this substance in the American medical literature since it has not only a strong bactericidal effect on most types of pneumococci in vitro, but can cure and protect animals from infections with virulent pneumococci."[119]

Florey assumed in his proposal, incorrectly at this early stage, that Dubos' substance was "an enzyme belonging probably to the same group of enzymes as [Fleming's] lysozyme," which Chain and others had recently purified and showed its power in dissolving bacterial cell walls. After Chain had assembled two hundred references to antibacterial substances during a literature survey, he and Florey decided around 1938 to make a joint investigation of microbial antagonism. By 1939, they were narrowing their search to Fleming's lysozyme and penicillin as well as the enzyme pyocyanase. Ironically, Chain assumed penicillin was also an enzyme and was more curious about the biochemical aspects of its instability, admitting that "It never struck me for a fraction of a second that this substance could be used from the chemotherapeutic point of view."[120] In contrast, Florey was motivated by the therapeutic value in the Dubos agent, so he added in the proposal that "the properties of penicillin, which are similar to those of lysozyme, hold out promise of its finding a practical application in the treatment of staphylococcal infections."[121]

Florey also cited Dubos in a November 1939 application to The Rockefeller Foundation (a separate organization from The Rockefeller Institute). He referred to the Dubos substance as evidence for "the possibly great practical significance of antagonistic substances produced by bacteria against bacteria." The Oxford team selected penicillin because it could be easily prepared in large quantities and was nontoxic to animals. The aim, as Florey stated in the application, was "obtaining in purified state and suitable for intravenous injection bacteriolytic and bactericidal substances against various kinds of pathogenic micro-organisms."[122]

In view of conflicting historical claims about when the penicillin research actually began, relative to Dubos' discovery and the extent of his influence on the penicillin work, Florey biographer Trevor I. Williams concluded that "all the evidence indicates that no experimental work on penicillin was done in Oxford before the late summer of 1939." He added that the diaries of Norman Heatley, Florey's young associate, confirm that full-time work on the project commenced around the beginning of October 1939.[123] The penicillin team itself reported that "it was not until late in 1939 that work on penicillin was taken up vigorously by Chain, Florey, and Heatley."[124]

Penicillin as a therapeutic agent was first reported in August 1940, a year and a half after Dubos' first announcement about gramicidin. The English team gave details for producing, extracting, purifying, and testing penicillin, with a prominent credit to Dubos. In August 1941, they reported their first clinical trial of penicillin in six patients at Radcliffe Infirmary, which established penicillin as a powerful and safe cure for bacterial infections. Penicillin, of course, became *the* nontoxic wonder drug and within a few years was an antibiotic of unrivaled efficacy around the world.[125]

Dubos' influence on streptomycin development. The gramicidin discovery sparked other scientists to probe the soil for bacteria that would produce more antibiotics. In 1940, Dubos' mentor Selman Waksman undertook several studies and eventually isolated seventeen antibiotics, of which four were generally practical. Streptomycin was the most important because it was the first drug isolated, developed, and used to treat human tuberculosis. As early as 1923, Waksman and Starkey observed that certain actinomycetes produced substances toxic to bacteria. Nearly a decade later, in 1932, Fred Beaudette, the Rutgers poultry pathologist who had been so helpful to Dubos, gave Waksman a culture of an avian strain of the tuberculosis organism *Mycobacterium*, whose growth was being inhibited by a mold. At the time, Waksman had a grant from the National Research Council to study the tubercle bacillus in soil and water, but the active work was undertaken by Waksman's student Chester Rhines.[126] Waksman admitted that in those days "The time was not ripe . . . to undertake extensive studies of antagonistic phenomena," and he was not yet thinking about producing selective toxic substances that contributed to therapeutics.[127]

Several months after the Dubos discovery, the Committee on Medical Research of the National Tuberculosis Association asked Waksman to search for soil microbes antagonistic to the tubercle bacillus. The minutes of a meeting on 2 April 1940, with Sharp & Dohme and Merck & Company, noted that Dubos had discovered a substance that destroyed gram-positive organisms. "Since the tubercle bacillus is also gram positive, there is a faint possibility that a similar effect might be exerted on the acid-fast [staining reaction for tubercle] bacilli. . . . If the Dubos substance has no effect on the acid-fast group, there is a possibility that other soil microorganisms, fungi, coccidia, et cetera, might exercise such an antagonistic function on the acid-fast group."[128] Waksman proposed at this meeting using the soil enrichment technique, as Dubos had done, to train soil organisms to produce antibacterial substances. Soon afterward, his laboratory began isolating agents from fungi and actinomycetes. In 1943, Beaudette brought them yet another odd culture, this time an organism grown from the throat of a chicken. Through collaborative research, Albert Schatz, Elizabeth Bugie, and Waksman recognized in the culture a strain of actinomycetes that he had isolated from soil three decades earlier. The chicken, Waksman recounted, "picked it up, no doubt, on a blade of grass from the ground on which it was feeding . . . this was the first isolation of the streptomycin-producing culture."[129] *Streptomyces griseus* was the organism that yielded streptomycin, the antibiotic introduced in 1944 that made Waksman famous.[130]

Several times Waksman underscored the influence of his student on his own work and on the antibiotic era. His autobiography, *My Life with Microbes*, credited the gold rush for antibiotics to Dubos's isolation of gramicidin because "to obtain the desired results required an analytical mind, an original coordination of all the facts, and especially a new philosophy . . . it was the beginning of an epoch."[131] In 1948, Dubos and Waksman shared the Albert Lasker Medical Research Award for their discoveries of soil antibiotics.[132]

In assessing Dubos' role in the development of antibiotics, it is not an overstatement to claim he discovered the first complete antibiotic. The fullness of his rationale and strategy involved several steps and a fusion of all the knowledge then available—from its isolation based on a natural antagonism, its development into a pure form, an understanding of how it worked, its immediate commercial production using the techniques and standards he and Hotchkiss established, and its ample and enduring use in the clinic. In essence, this research taught medical science the principles of finding and producing antibiotics and opened an interdisciplinary approach that drew on microbiology, chemistry, pharmacology, pathology, and medicine. While never a major contender like penicillin, gramicidin and tyrothricin formed the cornerstone of the antibiotic arsenal.

Dubos did not receive a Nobel Prize for this discovery, although he was nominated four times before 1952.[133] However, Domagk was recognized with the Prize in 1939 for the sulfonamides; Florey, Chain, and Fleming in 1945 for penicillin; and Waksman in 1952 for streptomycin. Given the confidential nature of Nobel committee deliberations, it may never be understood why Dubos was not so honored. Historically, the Prize is known for not always having been right, fair, or exhaustive, especially in recognizing theoretical breakthroughs over specific practical applications. Avery was also not awarded a Nobel Prize for his 1944 discovery of DNA as the genetic material, although he was nominated eight times between 1944 and 1952. The Nobel committee later regretted their error of omission, calling Avery's work "one of the most important achievements in genetics."[134]

Even so, the phenomenon of launching antibacterial therapy remains a biological landmark. Dubos' model work was basic to the penicillin and streptomycin efforts. It initiated the kind of scientific revolution that, in Thomas Kuhn's words, caused succeeding scientists to "work in a different world."[135] Many years later, Florey was bothered enough about Dubos not winning the Nobel Prize that he remarked to his colleague George Mackaness, "There really are only two things that sadden me over the penicillin story. One is the widespread, but erroneous, belief that much rancor and ill will existed within the penicillin team; and the other was the failure of the Nobel Prize Committee to recognize René Dubos for his most imaginative and pioneering work on antibiosis."[136]

Widespread praise and recognition did come to Dubos for his antibiotic discoveries, not just immediately but all through his life. His story was retold in great detail during the 1940s, in major magazines such as *Harper's*, *Fortune*, *Scientific American*, and in chapters of science books for the general public.[137] Whether the discovery was presented as a shrewd detective tale or a romantic adventure in science, its story of a wondrous cure captured the public's interest in a soil scientist who solved the scourge of infectious diseases. Other immediate awards from preeminent medical societies, peers, and academe included the 1940 John Phillips Memorial Award from the American College of Physicians, the 1941 E. Mead Johnson Award from the American Academy of Pediatrics, election in 1941, at age forty, to the National Academy of Sciences, and the first of forty-one honorary degrees bestowed on him during the following four decades.

A most meaningful recognition and honor came from The Rockefeller Institute in 1941 when he was appointed a full member and could establish a laboratory devoted to his own interests. This promotion was not an easy or automatic process, as Hospital chief Thomas Rivers recounted, because "A membership in the Institute is far better than a professorship in a university. You had to work hard to get it, and it was hard to come by, because the Board of Scientific Directors deliberately made it that way. Boy, you had to be something."[138]

The Hectic Glow of Success

At the midpoint of his life, Dubos was considered a premier scientist at the peak of his laboratory career, yet beneath the scientific triumphs lay the tragic irony that he had failed to discover a cure for the very diseases that afflicted Marie Louise. She was confined to bed in the Rockefeller Hospital from the time of his first announcement of the gramicidin discovery in February 1939, through the exciting moments of its publicity after the International Congress in September, until she moved that December to a sanatorium.

Marie Louise spent more than two years at the New York State Hospital for Incipient Tuberculosis in Ray Brook, located in the Adirondack Mountains between Saranac Lake and Lake Placid. Unlike the plush sanatorium a few miles away in Saranac Lake, Ray Brook was created by the state in 1900. It grew from a tent city with plank walkways, reflecting a burgeoning need to care for tuberculosis patients, into a sprawling complex of bland brick buildings. American poet Sylvia Plath, after a visit to her ill boyfriend in 1952, described Ray Brook's liver-colored interior walls, dark woodwork, and shoddy furniture as having a "funereal air."[139]

During this time, Dubos traveled back and forth to spend weekends with her. These visits were made more comfortable by Dr. Bray, superintendent of Ray Brook, who had sheltered Avery a few years earlier. Bray involved Dubos in research on tuberculosis in the sanatorium's laboratory and invited him to join the medical conferences and meals with the attending physicians.[140] He thus acquired a lengthy introduction to the rudimentary research being done on tuberculosis. Another scientist in the sanatorium laboratory was Keith Porter, a young Rockefeller cell biologist who was a tuberculosis patient (along with his wife Elizabeth) at Ray Brook. Porter long remembered Marie Louise's special difficulties in breathing that made her recovery impossible.

While Marie Louise's tuberculosis was not that severe, it was undoubtedly aggravated by the stress on her damaged heart. In a rare surviving letter to René, she wrote, "I must learn to become calmer and to master my irritability before getting much better."[141] Photographs taken on the sanatorium grounds show the Duboses in raccoon coats, warm hats, and heavy boots, suggesting they enjoyed walking in the cold mountain air. Their Saranac friends the Vorwalds remembered how they "even bob-sledded down Mt. Van Hovenberg at 60 mi per hr."[142] In all the pictures, Marie Louise appears pensive but sturdy and quite unlike patients with chronic wasting symptoms so typical of tuberculosis patients.

While preparing for her return home in 1942, Dubos accepted a position at Harvard University, in part to improve the environment for her continued rest. A few days before leaving Ray Brook, she wrote René, "Dr. Bray told me last Sunday, 'You should do well in the city, provided you do not let anything worry you.' Since my biggest task is to regain my strength, still so limited, I will try more than ever to not get upset about anything, absolutely anything. Anxiety exaggerates the difficulties, obscures the clarity of the spirit, and paralyzes the will and joy of living."[143] On returning to New York in time to celebrate Easter, the stresses of urban life quickly oppressed her. One day while walking beneath a music studio near Carnegie Hall, she panicked when she realized she was not strong enough to resume playing the piano or take up her usual activities. She also continued to worry about the welfare of her sisters in war-torn Europe. Within a month, Marie Louise died suddenly, on 24 April 1942 in New York Hospital. The death certificate stated only that the cause of death was not infectious without being more specific, but it clearly can be attributed to her weakened lungs and impaired heart.[144]

Searching for some consolation after Marie Louise's death brought René into a realm of despair and loneliness that never surfaced when he was around other people. A unique, poignant moment came a year after that April evening of her death when he wrote in French what for him was a flowery, stilted elegy. In it, he grappled with the symbol of spring that had somehow united them:

> Spring is an immense promise: each blade of grass that shoots up, each bud that swells open, each animal that carries numerous germinating cells has the power to find somewhere on earth another cell which will complete it and allow it to take root, and the power to become in its time a tree spreading out in the sunlight. . . . But why do so many germinating cells remain without being found, without becoming fruitful. And how many, even fruitful ones, remain barren to die of uneasiness on earth. . . . Marie Louise wanted to know only the springtime of life. . . . She loved the universe and expected everything from life.[145]

He went to Boston suffering from the shock of Marie Louise's death. Soon afterward, he wrote Avery an especially warm letter of appreciation for his fifteen years of friendship and guidance. In that letter, he announced "I am destroying or burying most of my New York past," and lamented the future would "never give me again the sensation of growth and discovery." Feeling like an exile in Boston, he added, "I have lost too much in too short a time to be able to make a rapid adjustment to my new environment."[146] From this time forward, he discarded all his personal correspondence and laboratory notes, made very few close friends, and adopted an inward lifestyle of the most intense privacy.

Developing a Wartime Vaccine against Dysentery

The two-year interlude that Dubos spent at Harvard was a static period of no growth and little change. For the young widower, it provided seclusion, peace, and a time for healing loss and grief. For the scientist, research in medical microbiology was limited to wartime needs using established techniques to produce other antibacterial therapies and a vaccine.

Harvard University welcomed Dubos to its faculty by awarding him an honorary degree on his arrival in Cambridge in June 1942. For the next two years he served as the George Fabyan Professor of Comparative Pathology and Tropical Medicine in the School of Public Health. The appointment suited his temperament. As the head of a small research department, with minimal administrative or teaching responsibilities, he could work alone or with a few quiet colleagues.

While his acceptance letter had stated a desire to study the physiology and immunology of tuberculosis, wartime military needs took precedence so that he, like hundreds of American microbiologists, embarked on projects proposed by the government. His defense-oriented project to produce a vaccine against bacillary dysentery began at Rockefeller (along with similar projects by his Institute colleagues) and continued after he moved to Harvard.[147]

All aspects of these war-related projects are difficult to document or explain fully. Part of the problem is that most material about biological warfare programs during World War II was destroyed after 1972 when the United States signed the Biological and Toxin Weapons Convention. However, a lot is known about projects of many scientists, including Dubos, because they were based on their current research interests; the studies were carried out in their own laboratories and their results were published in peer-reviewed scientific journals.[148]

Dubos was involved in two wartime programs, one as an advisor to the War Research Service (WRS) and the other as a contract laboratory researcher for the Office of Scientific Research and Development (OSRD).

The WRS program was established by the federal government and the National Academy of Sciences in 1942 to supervise twenty-six basic research projects about how bacterial warfare might affect crops, livestock, and human beings.[149] For the WRS, Dubos was one of many contributors of background notes—his were about the ecology and cultivating techniques for gram-negative pathogens, dysentery, plague, cholera, typhoid, and paratyphoid—used by a committee headed by Edwin B. Fred that recommended in 1942 that the War Department establish a biological warfare program. Dubos later acknowledged that his notes were not based on experience with these organisms but came from culling textbooks.[150] His slight familiarity with *Shigella dysenteriae* was acquired while he had been trying to learn why toxic substances secreted by gram-negative bacilli kept gramicidin from working. After the WRS program was established, he continued as a consultant on the *Shigella* portion, given the code name Project Y, until the Chemical Warfare Service replaced the WRS in 1944.

Most of Dubos' efforts were with the OSRD, headed by Vannevar Bush, which issued more than five hundred research contracts encompassing most military aspects of surgery and medicine, including tropical diseases. The OSRD initiated studies to find vaccines for nine pathogens—including anthrax and dysentery. Seven research contracts were awarded for various aspects of *Shigella dysenteriae*, including one to Dubos to look for antigenic elements in the Shiga group of bacilli that produce the exotoxin.[151]

In May 1942, he formally applied to the OSRD for a project on "immunochemical investigations directed toward active immunization and production of theraputic agents against bacillary dysentery," effectively making his war research an extension of his immediate interests and applying principles that had led successfully to isolating gramicidin.[152] After the contract was transferred to Harvard, his laboratory quickly devised two ways to grow shigella in sufficient quantity for experimentation. Both were nontraditional means to force air into the culture: by violently agitating the liquid medium;[153] or by circulating the medium over the bacteria against a counterflow of air, a technique similar to one described for large-scale production of penicillin.[154] Dubos then concluded he had "reached the end of what can be done on the scale of ordinary laboratory operations."[155]

Despite their limited output of organisms, the laboratory also produced two potential therapies against dysentery. The first one reported to OSRD on 19 January 1943 was "able to immunize white mice against bacillary dysentery infection by treating them with a saprophytic organism unrelated to dysentery." However, the committee denied Dubos permission to discuss or publish information on this finding, because "this entire work is in a very preliminary stage."[156] The work apparently was not pursued.

Experiments with another intuitive but odd therapy were published in *The Journal of Experimental Medicine* in September 1943.[157] This treatment used bacteriophages, the viruses that prey on *Shigella*. Although not entirely innovative, the treatment was clinically attractive. It had first been used by Felix d'Herelle in the early twentieth century to eliminate dysentery in humans.[158] A significant drawback for using phages was that they, like antibiotics, select variant forms that foster dangerous resistant populations of bacteria. Aware of this problem, Dubos created a different method to deliver enormous amounts of bacteriophages to infected animals via the tail, reasoning this would overwhelm 99.9 percent of the bacteria and would leave too few pathogens to establish new infection, prevent mutation, and avoid resistance.

Despite overwhelming success with this therapy in mice, no further tests were made to establish whether such large doses were toxic or how the technique would translate to human medicine.[159] Of all the contributions to this dysentery project, Dubos was enamored by implications of phage therapy and later thought it was the only aspect worth further exploration. The laboratory's work on therapies was soon replaced by research to create a vaccine against the disease.

The fear that *Shigella* would be used as a biological weapon during the war was undoubtedly eased after Dubos advised the OSRD that his laboratory had made "a number of immunological studies on sera obtained from German, Italian, and Japanese prisoners of war."[160] The tests showed that the prisoners had no antibodies against dysentery, providing good evidence that these countries were not vaccinating their troops against dysentery. These tests also led to an assay to measure the efficacy of the experimental dysentery vaccine.

Subsequently, his laboratory worked out the practical aspects of preparing and testing a vaccine.[161] Since even larger quantities of bacilli were needed and the laboratory was poorly equipped to produce them, Dubos asked the OSRD's Committee on Medical Research to arrange with E.R. Squibb & Sons for the commercial production of a toxoid (a nontoxic but immunogenic derivative of the exotoxin).[162] As a result, maximum toxin production from Shiga bacilli was achieved by eliminating iron from the culture medium, and a safer way was found to produce the toxin by using an avirulent Shiga strain. Finally, the toxin was detoxified with formaldehyde, a technique that had been developed by other OSRD scientists to produce toxoid vaccines against botulinum and diphtheria. The Shiga toxoid vaccine successfully protected mice against exposure to either high doses of toxin or of virulent bacilli.

Fortunately, bacillary dysentery did not become a great threat during the war and sulfa drugs were used to treat the occasional cases of dysentery that occurred in military camps. Squibb ended its production of this vaccine before assessing its toxicity in humans.[163] The dysentery project ended in 1946 with Dubos' final reports to the OSRD and Rockefeller's Board of Scientific Directors, along with a publication in *The Journal of Experimental Medicine* describing techniques for preparing the vaccine.[164] In 1955, he observed that "the nature of my scientific research during the war was not of very great interest. . . . It was just one of the things that one does during war."[165] This comment fits with an underlying frustration that he had not added any new theoretical knowledge about bacteria or antibacterial therapies during these years.

During collaborations with Avery in medical microbiology, Dubos gained and contributed to a broadened view of infectious diseases. His experience contrasted with those researchers, following Pasteur during the so-called Golden Age of Bacteriology, who limited their experiments to bacteria alone as they worked under the influence of the germ theory. Dubos opened new realms of theoretical bacteriology when he took an ecological rather than the laboratory's more chemical approach to solving Avery's problem. Traditionally, when medicines were found by empirical methods, scientists would then hunt and assign causes for how they worked. Avery's serum therapy, for example, led backward to learning how an antibody effectively reacts with the capsule in virulent pneumococci. Dubos' discovery of the S III enzyme that dissolved the capsule further extended this reverse, reductionistic process of discovery.

Then, forging an independent course, he turned this process around and put experimental medicine on a forward, more rational basis. His uncommon grasp of the interplay of bacteria with their environments—in soil samples, laboratory cultures, and animal tissues—guided the search for new medicines. He exploited these bacterial antagonisms to isolate their antibiotic substances. The importance and novelty of this research was the successful coordination of bacteriological, chemical, and clinical research that influenced researchers worldwide, leading to a scientific medicine that gave us the antibiotic era. His immediate discovery, however, produced the less than perfect gramicidin, which was successful on its own but fell short of the crowning achievement he expected.

Simultaneously, by fathoming other levels of microbial ecology, he was becoming aware of why infectious diseases were more complex and challenging. The pneumococcus and streptococcus are not single well-defined species but instead multiple types, each type significant in the degree of disease it causes. These differences made diseases more difficult to treat, so that a one-germ one-chemical solution was not reasonable: serum therapy had to be tailored against each type of pneumococcus; sulfa drugs worked against some streptococci but not those that caused rheumatic fever; gramicidin cured local infections but was unsafe against systemic infections. In addition, therapy did not work when sick patients and animals had lowered resistance or were in poor health.

Rudimentary as these observations may seem today, they point to some universal aspects of microbes. Years before the antibiotic era blossomed, Dubos was among the earliest to perceive problems posed by bacteria: their adaptability and versatility are potential sources of drug failure, resistance, and even the cause of previously unknown diseases. More critical, the physiological condition of the patient influences the activity of the medicine, the pathogens, and the course of the disease. It no longer seemed enough to study microbial activities, because the patient and the environment are equal variables in the disease process—ideas that are now virtually axiomatic although then at the forefront of an upheaval in experimental medicine.

Without a clear vision of how to proceed at the end of the war, Dubos was restless to embark in another scientific direction. The stage was now set for the bacteriologist—trained in ways of pathogenic microbes, partial to their interactions with their diseased host, and cautious about the value of antibiotics in general—to tackle tuberculosis, a disease with a unique set of bacterial and environmental interactions.

3 Tuberculosis and Dilemmas of Modern Medicine

Your remedy does not treat the real seat of evil. It continually removes the traces of the enemy but it still leaves him deep in the invaded country.

ARTHUR CONAN DOYLE, 1890,
on Koch's vaccine against tuberculosis

DUBOS RETURNED TO THE ROCKEFELLER INSTITUTE in July 1944, just after submitting the manuscript for his first book, *The Bacterial Cell*, to Harvard University Press. Accepting his only reason for returning would be deceptively easy. Although he experienced a peculiar emptiness in Boston that "was associated in my mind with the loss of my first wife," he was more bereft than this statement reveals.[1] He was also cut off from relatives in war-torn France until 1944 when he learned that his brother was a German prisoner of war and his letter about Marie Louise's death reached them.

At Harvard, he encountered many "picturesque, odd, unique, interesting human beings" from all areas of the humanities and sciences who made his tenure "by far the richest I've ever spent."[2] Yet, after only a year, Dubos wrote Thomas Rivers, chief of the Hospital, asking to return. Rivers had warned him on leaving that "you are going to be fed up" with administrative duties and social responsibilities that come with being "a professor at a university. However, when you are ready to come back to the Institute, just call me and I'll talk to Dr. Gasser. I am sure that he will be glad to have you back."[3] Dubos reminded Rivers of this offer and asked whether it would be a wise step, because he was formulating "more precisely my own plans for the future."[4] Within three months, Gasser and the Board of Scientific Directors approved a salary equal to what Harvard was paying, $12,000, as well as the considerable sum of $10,700 to furnish a tuberculosis laboratory in New North Building (now Theobald Smith Hall).[5]

The move was uncommonly abrupt but, as the letter suggested, Dubos was giving serious thought to studies of tuberculosis. Some reasons for returning were those Rivers had foreseen. Harvard wanted to reorganize the teaching of tropical medicine in his department. Although other options were extended, their future plans were incompatible with his lukewarm interest in classroom teaching and an even greater lack of talent and enthusiasm for managing an academic department. Such distaste for administrative duties as well as his desire to avoid stressful situations and controversies kept him from accepting appointments at many universities over the ensuing thirty-five years, including directorship of the Pasteur Institute.[6]

A more personal reason for leaving was implied in a parable he composed for A. Baird Hastings, a former colleague at Harvard, on the occasion of Hastings' fiftieth birthday in 1945. It begins, "a young soldier returned from war came to the Learned Man asking for the formula of Wisdom and Happiness," which had been revealed to him by the "Goddess of Power in the deep and mysterious temples of the OSRD." The Learned Man told the soldier to avoid that kind of "grey knowledge" for "it leaves you powerless against your greatest enemy, against yourself." The soldier was then given two pieces of advice: "worship in the luminous Gardens of Biochemistry and Physiology where you will hear the harmonious concert of healthy normal life"; and "Know Thyself . . . then watching without despair the eternal folly of men, you will be amused by their various antics and will even enjoy participating in them." Hastings later observed that this parable "describes René's secret of living."[7]

The more openly expressed motives for returning to Rockefeller were to assert research independence and to make a fresh start in bacteriology on a disease that most peers at Harvard would have likely considered scientifically *passé*. Tuberculosis did not fit in departments of tropical disease, pathology, or even bacteriology. At the time, the disease was neither a public health threat, nor a pressing research problem, nor as glamorous as the newest research on viral diseases.

Dubos had serious ecological concerns about this ancient disease. In the 1940s it was the seventh leading cause of death in the United States and the most common cause of death between the ages of fifteen and forty-five. Although the disease seemed under control, the virulent microbes remained constantly present in the population, and they were more poorly understood than those of any other infectious disease. Arnold R. Rich, in his classic text *The Pathogenesis of Tuberculosis* (1944), criticized several prominent students of tuberculosis and epidemiology who were concluding "tuberculosis is now definitely on the way toward eradication." Rich called this "premature optimism" because it "takes too little account of the prevalence of 'sub-clinical' or 'carrier' infections in tuberculosis."[8]

Dubos was known for claiming the most important advances come from individuals who stray from obvious paths and venture into unexplored areas. "It is not easy," he said, "to integrate these temperamental trailblazers into the rigid and cumbersome structure of large educational or research institutions."[9] If he felt personally lost or unneeded in the large scientific empire of Harvard, he clearly felt confident and free to blaze a new path at Rockefeller.

Before going to Harvard he had begun to question whether his theories and experimental approach had neglected some of the obscure reactions to infection. Then several things happened on his return, all of which introduced complexities into modern medicine. He was the first to predict dangers from microbial resistance to antibiotics. His searches for other antibiotics against tuberculosis were futile and abandoned. The initial experiments on the tubercle bacillus made rapid, ingenious progress but simultaneously dispelled even further the optimistic promise of antibiotics and vaccines. *The Bacterial Cell* exposed some of this turmoil as he wavered between gathering facts and producing theories and between knowing bacteria and understanding disease. The prescient skepticism of Dubos during these years forms the substance of chapters 3 and 4 and shows that much of what he recognized about disease continues to haunt medicine today.

Microbial Resistance to Antibiotics

As early as 1942, Dubos warned that bacterial resistance to antibiotics should be expected. His prediction was based in large part on almost twenty years of research on bacterial responses to environmental factors. He knew that "susceptible bacterial species often give rise by 'training' to variants endowed with great resistance to these agents. In some cases, drug resistance may be due to changes in metabolic behavior . . . [or] may result from a change in cell permeability."[10] What is impressive about this first warning of an impending crisis in medicine is how early it appeared, just as penicillin and gramicidin were emerging from laboratory tests and not yet in general use.

More significantly, his prediction is based on a doctrine of bacterial variability that was taking shape in the 1940s.[11] Until the mid-nineteenth century, microscopists believed bacteria were simply various forms of one or a very few species. Louis Pasteur and Ferdinand Cohn were among the rare scientists who rejected this doctrine called pleomorphism and believed instead in monomorphism, or the distinctiveness and stability of multiple bacterial species. Evidence of this distinctiveness was supported by many scientific forces over several decades, most positively by the methods developed by Robert Koch to isolate and grow microbes in pure culture. By the 1920s, however, a few bacteriologists, among them Sergei Winogradsky and Theobald Smith, were beginning to notice that this doctrine of monomorphism did not account for "the plasticity of microorganisms within certain limits." To account for the various forms of bacteria they observed within species, Avery and Dubos considered bacterial variability from the point of view of "training" or "adapting" cultures to new substances, new hosts, or new environmental conditions; other scientists, like Rockefeller's Paul de Kruif, used the term *microbial dissociation*, while still others borrowed terms of classical genetics like *mutations*, *genotypes*, and *phenotypes*.[12]

When Dubos wrote *The Bacterial Cell* he criticized germ theory bacteriologists for their "blind acceptance" of the doctrine of monomorphism because it had "discouraged for many years the study of the problems of morphology, inheritance, and variation in bacteria." In a chapter devoted to bacterial variability, he discussed numerous investigations giving evidence that bacterial species vary in form, function, and chemical and antigenic composition. Bacterial transformations, he concluded, whether they are "permanent or transient—not only of a quantitative but often of a qualitative nature—appear in an unpredictable manner under conditions where the 'purity' of the culture cannot be doubted."[13]

At that time, it was not known that bacteria have genes or that DNA is the genetic material. Dubos had no inkling of bacterial genetics. In fact, only when *The Bacterial Cell* was in page proof did he insert an addendum with fifty-three micrographs by Carl F. Robinow, who provided evidence from late 1944 that bacteria possess a discrete nuclear body.[14] While today's discussions of drug resistance focus on molecular mechanisms of bacterial genes that are mutated, recombined, or switched off and on, before 1945 there was no knowledge of this level of response.[15] Thus it happened that Dubos described the dangers of resistance in terms of several unrelated and unknown mechanisms of bacterial variability and adaptability.[16]

Before 1942, an empirical body of knowledge existed about drug resistance. Paul Ehrlich found resistance to arsenicals used to treat syphilis in 1907 and called the phenomenon "drug fastness." Alexander Fleming found resistance to the naturally occurring antiseptic lysozyme in 1922. In 1939, members of Avery's laboratory witnessed some of the earliest resistance to the sulfa drugs. Colin MacLeod was able to produce in vitro resistance to the drugs by "training" pneumococci, that is, by making a progressive selection of the few resistant forms normally growing in a susceptible culture. He also found that some Hospital patients failed to respond to serum therapy because the invading pneumococci were already resistant. When these patients were treated with both serum and sulfa drugs, this combined therapy was more effective than either agent alone.[17]

Based on these and other studies, the Avery laboratory saw firsthand how these agents were producing unpredictable responses. As a result, they cautioned that bacterial resistance occurs "not only in culture media, but in the bacteria themselves and in certain tissues and fluids of the animal body."[18] The drugs could be destroyed, neutralized, or transformed by other substances in the body; they could fail to diffuse in adequate quantities; or they could simply fail due to any number of unidentified causes. Whatever its cause or form, they were made aware that bacterial resistance should be anticipated. This experience with sulfa drugs was soon found to be applicable to gramicidin (described earlier) and penicillin.

Surprising as this seems now, warnings of bacterial resistance went unheeded for many years. Scientific discussions of antibiotics focused on the medical cures they produced. Incidents of resistance, when reported, were buried in medical scientific literature read by few laboratory scientists and fewer physicians. Reports of drug failures, adverse reactions, and bacterial resistance were eclipsed by the overwhelming triumphs of antibiotics. When measured against genuine, even miraculous, cures from previously fatal infectious diseases, the cases of resistance appeared statistically insignificant. However, resistant strains of microbes displaced susceptible ones, and superinfections (with new species of microbes) replaced original infections, making both conditions much more problematic and persistent. Nonetheless, most microbiologists and physicians at the time believed the germ theory of disease was doctrine and that diseases could be eliminated simply by removing the responsible microbes. To these believers, antibiotics seemed a definitive answer to eradicating infectious diseases.

Fortunately, Dubos was not content with a single early warning. Two major lectures in 1944, still before penicillin was available to civilians, contain more explicit warnings of resistance. One lecture surveyed the state of infectious diseases for an American Philosophical Society symposium, "Wartime Advances in Medicine." This time the warning of bacterial resistance dispelled the "dawn of a new era" in medicine by pointing out that microbes exhibit such extraordinary plasticity and variation in their biological properties that they would evade the new antibiotics by producing resistant strains. Human life had adapted and survived epidemics throughout history without benefit of scientific measures to combat these diseases. Environmental changes, he argued, had played a role in their disappearance and would play a role in their reemergence. Foretelling a path he himself would eventually take, he advocated a return to the "main channels of biological and biochemical philosophy" to consider environmental encounters where host and parasite converge.[19]

The other lecture, actually a series of eight called the Lowell Lectures, formed chapters of *The Bacterial Cell.* While the book reflects his research and study of bacteria under Avery's influence, its initial chapters are a compendium of facts on structures and functions in bacteria as well as techniques for handling them. The following chapters introduce some bold perspectives on the influence of the environment on bacterial variability, virulence, immunity, and chemotherapy. Joshua Lederberg, bacterial geneticist and Nobel laureate, who was a medical student when the book was published, has called it a "book that only Dubos could have written, because he brings into sharp focus what we now take for granted—namely, that bacteria are cells."[20]

The Bacterial Cell provides experimental evidence from other scientists to show the various ways in which antibiotic resistance can occur. In all cases, some bacteria still thrive in their altered state and behave differently. So many properties and characters of bacteria can undergo independent variation, Dubos warned, that "drug fastness can occur without any alteration of virulence or can be associated with structural or metabolic variations which affect one or several of the factors of virulence."[21] He was the first to foresee that the new medicines were introducing unknown perils into an evolutionary pathway of bacteria, a warning that has become a terrible reality: microbial worlds altered by antibiotics have and will continue to determine the diseases that affect us.

More personal, though less obvious, this book frames a changing attitude toward research. Reflecting the sentiment in the book's epigraph by nineteenth-century French physiologist and "father" of experimental medicine, Claude Bernard, Dubos was emulating a "man of research" who concentrates on finite facts while justifying a philosophical spirit to discuss problems of science. In the final chapter, he changed and adopted Bernard's complementary attitude of a "man of science," literally, a man of knowledge, who creates a pattern or imposes an order on the confusing accretion of facts. From this latter stance, he criticized the theories of biochemical specificity and specific etiology of infectious diseases as the kind of experimental medicine that may have seemed convenient and logical but was instead producing "discoveries in reverse" that could not account for "the logic of life." He further criticized these traditional investigators for being attracted "by the dynamic rather than by the intellectual aspects of the new problems." The book ends with a partial rejection of the biochemical explanation of life that he had embraced on entering the Institute in 1927. He quotes Bernard's view that "It is no chance encounter of physico-chemical phenomena which constructs each being according to a pre-existing plan, and produces the admirable subordination and the *harmo-*

nious concert of organic activity. There is an arrangement in the living being, a kind of regulated activity, which must never be neglected, because it is in truth the most striking characteristic of living beings" (emphasis added). The real challenges in science, Dubos predicted, lay in digesting and integrating the facts to penetrate the "harmonious concert of organic activity."[22]

The Bacterial Cell effectively closed down an old science of looking at bacteria as objects and envisioned other sciences that integrate bacteria as cells in living processes. According to Lederberg, the monograph contained "the seeds of its own obsolescence since it inspired so many to pick up on his inspirations and challenges. In doing so, they brought about a rapid displacement of what Dubos said and substantially furthered the march of bacteriology."[23] This is the moment when Dubos took a separate scientific path from the bacterial geneticists and molecular biologists who revolutionized the field in one direction with their focus on the machinery of parasites. In these pages, one senses an introspective humanistic scientist emerging who would revolutionize bacteriology in another direction with a focus on the relationships between host and pathogen in the disease process.[24]

Establishing a Tuberculosis Laboratory

When Dubos returned to Rockefeller to organize a tuberculosis laboratory, three members of his Harvard laboratory came with him: Gardner Middlebrook, a young physician specializing in tuberculosis, and the two researchers who had gone with him to Boston, Cynthia Pierce and Letha Jean Porter. Jean was credited in *The Bacterial Cell* as the Miss Porter who "read the whole manuscript in an attempt to conceal the foreign flavor of my English style." In October 1946, Jean became his second wife and continued to play an active role in all his subsequent writings.

During the next twenty-seven years, until his retirement in 1971, fifty scientists and physicians from all over the world joined the laboratory, with appointments ranging from a few months to several years. Four became members and professors: Rollin D. Hotchkiss, Merrill W. Chase, James G. Hirsch, and Zanvil A. Cohn.

The young scientists initially encountered a scientific milieu that was not unlike the one Dubos found on entering Oswald Avery's laboratory in 1927. There was no formal indoctrination or training. Newcomers were left to find their own way, causing them to complain about a "cold shoulder treatment" that somehow was not bringing them into the laboratory's major activities. Dubos, however, viewed this as an environment free of constraints that would foster curious investigators and not mere problem solvers. Problems and experiments were not assigned, so a newcomer had to find a project suited to his own taste and gifts. The atmosphere required secure people with strong inner direction. Once an individual found a suitable project, Dubos became a constant source of counsel, gently maneuvering the newcomer into the group.

This style effectively "trained" many illustrious medical scientists who remained productive wherever they went. One postdoctoral fellow was Bernard D. Davis, physician and bacterial geneticist, who said that during his year in the laboratory, "we argued incessantly and intensely about all kinds of problems, scientific and social. I was the naive, idealistic young American, seeking absolute truths and social perfection, while he was the worldly European, seeing subtle complexities in every problem. We were rather like Herr Settembrini and Herr Naptha in *The Magic Mountain.*" What they discussed were always scientific ideas, not personal ones, for "beneath his showmanship was a great deal of reserve."[25]

After a workday dedicated to experiments, colleagues gathered in Dubos' office. There, with feet on his desk and hands folded behind his head or pulling on wisps of hair, Dubos would speculate on what the day's results suggested. Virologist Frank Fenner recalled that "any exciting lead formed the base of an inverted pyramid of heady speculation, which often as not collapsed the next day. But we were all, most of all Dubos, enthusiastic about the work, and we were stimulated vastly by the imaginative leaps that René made."[26]

A daring rationale. Tuberculosis posed first-rate research problems, whether they concerned acute and chronic phases of disease, silent infection versus overt disease, or aspects of virulence and immunity. The disease is greatly influenced by environmental factors such as poor nutrition, exhausting work, and crowded living or working conditions. Its microbes are so widespread in the population that hidden infections erupt into epidemics in time of war, famine, and other sudden social and economic disruptions. Many aspects of these pathogenic bacilli interacting with human hosts were not and still are not understood.[27]

When this investigation began, it seemed more knowledge of the microbe was not needed and that the disease was adequately under control. The causative organism, *Mycobacterium tuberculosis*, had been discovered in 1882 by Robert Koch. Albert Calmette and Camille Guérin introduced the BCG vaccine against tuberculosis in 1921. The tuberculin test and X rays were in use as the main tools of detection. Many sanitary measures and public health controls, including sanatoria that quarantined patients, were actively containing its spread. In 1944 the Waksman laboratory isolated streptomycin, an antibiotic that initially promised a cure. All these factors contributed to declining death rates worldwide. In the United States from 1900 to 1950, deaths from tuberculosis decreased from two hundred to twenty per one hundred thousand population.

Nonetheless, Dubos found a pressing need to study a disease that was a still major killer and for which there was no complete cure. Alarmingly, new tuberculous infections were still increasing, even with a vaccine, which meant there was a reservoir of contagion creeping silently through populations. Scientists who predicted that tuberculosis could be eradicated were overlooking the fact that this goal depended not only on rigorous medical therapy but on a relentless endeavor of early detection, segregation, and education to prevent infection and halt the spread of the contagious pathogen.

In addition to these social factors, Dubos had both personal and professional reasons for studying tuberculosis. Personally, he was puzzled why Marie Louise should have died from it since they lived in a protected suburban environment where she was not exposed to the disease. As discussed earlier, she had incurred a mild tubercular infection as a child and her symptoms had disappeared years before. Dubos speculated the disease reappeared due to stresses brought on by her frail constitution, her sister's mental breakdown, and the wartime separation from her French family, all of which impaired the mechanisms that were keeping her disease dormant. Her experiences alerted him not just to a balance between humans and bacteria but especially to the impact of physical and mental environments on that balance.

From a professional perspective, there were other reasons to learn how a mild infection could turn into overt disease. Tuberculosis behaved differently than the acute cases of pneumonia, meningitis, and bacteremia studied in the Rockefeller Hospital. It was a chronic illness, comparable to rheumatic fever, and involved recurrent bouts, remissions, and lapses, and ended in fulminating illness and agonizing death. Recalling Winogradsky's emphasis on microbial interactions, Dubos perceived that a tuberculous infection teemed with interactions as complex as those found in soil. There were superb riddles posed by the dynamics of health and disease in tuberculosis: the pathogens had fluctuating potentials and the hosts exhibited wayward sensitivity.

A hazardous organism. Undertaking research on tubercle bacilli was a formidable task. A few facts about this bacillus show why little progress had been made in the half century since Koch discovered it. *Mycobacterium tuberculosis* is a rod-shaped bacillus that is notable for its thick waxy cell wall, simple diet, and slow growth. This organism is aerobic, which means it reproduces best in tissues such as the lung that are rich in oxygen. It has the ability both to persist in the body and to spread easily from person to person. Under optimal laboratory conditions, this bacillus requires up to twenty-four hours to undergo one cycle of replication, while most bacterial species replicate within several minutes. It takes nearly two weeks for a visible colony of mycobacteria to appear on a solid culture medium. Until 1946 the tempo of working with this organism was such that a single experiment, from growing the inoculum to preparing extracts to infecting mice, entailed a lapsed time of approximately four to five months. There were of course overlapping experiments, but impatience with long periods of waiting for results led to work on other species of bacteria and problems that could be answered more rapidly.

Laboratory studies of tuberculosis were not only enormously time consuming. They were dangerous. Few scientists wanted to risk culturing and infecting animals with a virulent organism that is so readily aerosolized and contagious. The specialized biocontainment facilities that are required today did not exist then. Davis noted that most early researchers who worked on this bacillus got started while they were convalescing in a sanatorium where they acquired partial immunity. More frightfully, he added, "There was a strong medical tradition that laboratory investigators of infectious disease, like physicians in epidemics, must accept the risk of being exposed to dangerous agents as part of the job."[28]

The hazardous experiments in the Dubos laboratory were carried out on open laboratory benches. According to Davis, they "used hoods with ultraviolet radiation but no air flow, and we actually used unplugged pipettes to deal with virulent tubercle bacilli—I once got a mouthful of a culture." Many researchers developed clinical tuberculosis including Davis, Gardner Middlebrook, and Dubos' second wife, Jean. Dubos had likely acquired some immunity while working in the Ray Brook laboratory, for when he developed a tuberculoma in an arm after cutting it on a broken, contaminated test tube, the bacilli did not spread and the infection healed. As a result, Gasser insisted that a "safety suite," or rudimentary negative pressure isolation unit, be constructed on the lowest floor of the Institute's animal house.[29] This new space was then devoted to studying virulent tuberculosis. Because the unit was in another building, the main laboratory worked with nonpathogenic strains of mycobacteria, a murine (mouse) form of pseudotuberculosis called *Corynebacterium kutscheri*, and eventually other less dangerous pathogens, particularly staphylococci. Even today, despite biocontainment facilities, tubercle bacilli continue to pose great hurdles in the experimental study of tuberculosis.

Choosing a strategy. During visits to Ray Brook where Marie Louise was hospitalized, Dubos gained some working knowledge of how to handle the mycobacteria. He also benefitted from earlier work of Rockefeller scientists who had clearly demonstrated the importance of the environment on this disease.

Pathologist Theobald Smith had the most influence on Dubos' approach to tuberculosis research. In 1898, he detected the difference between the bacilli of human and bovine (cattle) tuberculosis, based on the distinctive types of disease they cause. As one of the Institute's original board members, he maintained that "the true function of medical science is to study the disturbances of equilibrium between man and his environment, to anticipate them and to suggest those compensatory movements which will counterbalance the temporary ill effects of social movements."[30] In his book *Parasitism and Disease* (1934), which Dubos studied closely, Smith introduced the ecological view that parasitism, or infection, is a normal phenomenon with a complex give and take, so that exaggerating any one facet may upset others and lead to disease. He also emphasized that the physiological status of the host was as important as the nature of the parasite for determining the severity and outcome of an infection.[31] In many experiments, Dubos tested factors that disturbed this equilibrium.

Other physician-scientists studied how a host's white blood cells and tissues respond to infection. Board member and pathologist T. Mitchell Prudden demonstrated in 1891 how monocytes enlarge after ingesting the bacilli and crowd together to form a tubercle, the characteristic lesion of the disease. Prudden observed in his book on bacteria for the general public that "the healthy body is not good soil for the tubercle bacillus."[32] During the 1920s, pathologist James B. Murphy showed that when leukocytes are destroyed by radiation from X rays, the hosts become much more susceptible to tuberculosis. Histologist Florence Sabin, the Institute's first female member, who retired in 1938, was the only one prior to Dubos to study tuberculosis exclusively. Her studies on cellular and immune reactions influenced many of Dubos' experiments, including those on chemical fractions of the bacillus that produce tuberculous lesions as well as those on the influence of nutrition and oxygen in the tissues on the progress of the disease. Tangible evidence of their long relationship came when Sabin presented him in 1951 with the highly regarded Trudeau Medal from the National Tuberculosis Association. In her cogent analyses of his new research concepts and innovative techniques, she praised his "keen, critical judgment" and recognized his work as "marking a new period in the study of this disease."[33]

Pathologist Eugene Opie studied tuberculosis in families and found evidence for its communicability within households and for high susceptibility to it among young adults who moved from rural areas to crowded cities. His studies of latent infections detected viable organisms in lung lesions. The vaccine he produced in 1939 used heat-killed tubercle bacilli and was successful in clinical trials. Fourteen years later, Dubos was encouraged to try other versions of a killed vaccine.[34]

These early scientists essentially focused on the host defense whereas Dubos was more familiar with biological and chemical studies of bacteria. Understandably, he chose to begin with what he knew best—to learn more about the pathogen—and notably this was something that had not yet been attempted at Rockefeller.

From 1944 to 1950 four topics were explored: antibiotics, culture techniques, detection tests, and vaccines. Some of these experiments now seem old or inconclusive, but for three decades they were considered major breakthroughs. From two discoveries came methods still in use today. Even for those that have not fared well, the scientific thinking behind the experiments provides critical insights into Dubos' enlarging perspectives on disease.

Antibacterial Agents against Tuberculosis

Despite claims that he ended his search for antibacterial agents with the discovery of gramicidin—because they were no longer intellectually challenging and more appropriate for pharmaceutical manufacturers—Dubos continued to look for antibiotics effective against the tubercle bacillus until 1948.

Unsuccessful efforts. In 1944 Dubos requested support from the OSRD (which was funding his shigella work) to find a drug against tuberculosis that would intervene in locally infected host tissues rather than attack the bacillus directly. He reasoned that while sulfa preparations were too basic to work in inflamed tissues, antibacterial substances with an acidic radical might work. The OSRD approved this ingenious ecological strategy for a "full scale attack on the chemotherapy of tuberculosis."[35]

During the next four years, as Dubos looked for candidate drugs among antimalarial compounds provided by the OSRD and synthetic acids containing plant hormones, the laboratory reported at least five substances in publications. The first of these, which emerged from research collaborations of Dubos and Porter around 1945, was an extract from the fungus *Aspergillus ustus* isolated from patients with aspergillosis at Ray Brook sanatorium. Despite its purification by Lyman Craig using his new countercurrent technique, its tests on mice were inconclusive. Two years later, Alfred Marshak and Cutting Favour reported antituberculosis activity of two extracts, one from lichens and another from tuberculin, but these were not subjected to animal tests. The final and most promising substance was a nontoxic sulfone called Equityl that had limited success in mice and humans.[36]

Productive investigations. A search for antibacterial agents produced naturally by the body began in 1950 when physician James G. Hirsch joined the laboratory. Hirsch soon isolated lysozyme, spermine, an unnamed thymus peptide, and one from leukocytes he named phagocytin.[37] Although he did not purify phagocytin, this previously unknown substance set off a chain of discoveries for natural host antibiotics, including what are today called "defensins." According to physician-scientist Paul Beeson, this innovative research by the Dubos laboratory represented "a significant direction for a better understanding of infection, namely, identification in various tissues of naturally occurring substances that either favor or retard the growth of pathogenic microorganisms."[38]

A serious deterrent to searches for more antibiotics was increasing bacterial resistance. In 1946, two years after streptomycin was introduced, profound resistance to it was found in patients with tuberculosis. Some of this research was done by Middlebrook in the Dubos laboratory as a collaboration with the Ray Brook scientists.[39] Dubos took up the warning against streptomycin, saying that in addition to causing nerve injury and impaired hearing, the drug was harmful because tubercle bacilli were quickly becoming resistant to streptomycin. When patients developed resistance to streptomycin, physicians added the antibiotic para-amino-salicylic acid (PAS), which delayed but did not prevent resistance.

Hirsch and Dubos were among the earliest observers of multi-drug resistance, which occurred during the laboratory's only clinical study of tuberculosis in the mid-1950s. They tested whether combining exercise with antibiotic therapy could eliminate enforced bed rest and concluded the regimen had no harmful effects on a patient's recovery. Nevertheless, the scientists carefully noted the high incidence of drug-resistant infections in their small trial group, and they rejected two of twenty-three candidates for the study because of "triple-drug resistance" to streptomycin, isoniazid, and PAS; five patients were resistant to one drug and one patient developed resistance during the study.[40]

This led the laboratory to devise a test, using the routinely collected sputum samples, to monitor whether patients had developed resistance to any or all of these drugs. Test results were available within a week, which was very rapid for the 1950s, so that the drug regimen could be revised promptly.[41]

Dubos declined co-authorship of this clinical study, because he was not a physician, even though he actively joined in rounds of the ward and X-ray analyses. Hirsch observed how Dubos loved to "play doctor," replacing his usual tan laboratory coat with a physician's white coat when visiting patients. Before this study was reported, Dubos had concluded that trying to quell resistance by using several drugs at once was both dangerous to the patient as well as confusing to the investigator since there were too many variables to analyze.

By 1948, the search for antibiotics ceased. In part, Dubos was too impatient to make endless systematic searches into thousands of compounds to find the rare, useful few. The fashion of searching for antibiotics, he remarked, had become the "favorite indoor sport of many biologists."[42] He was also critical of the weaknesses in empirical screening methods to gamble on finding a lucky one. Hotchkiss said he hardly raised his head to look when "I once showed him a paper on the second one thousand isolations of antibiotics from soil."[43] Davis, who witnessed this intellectual restlessness, said Dubos "expressed doubt that any antibiotic was likely to turn up again with the fortunate nontoxicity of penicillin; hence the major future advances against infectious diseases would have to arise from increased understanding of the organisms and the host defenses."[44]

It is clear from his own experiments that Dubos became sensitized to, and frustrated by, biochemical complexities associated with antibacterial substances in inflamed tissues. Antibiotics were creating a mirage of medicine, an illusion of what drugs could do, and, as he observed, they were not reaching diseases at their origin. Concurring with physician and detective-story writer Arthur Conan Doyle, he believed a remedy may remove "traces of the enemy but it still leaves him deep in the invaded country."[45]

Rapid and Dispersed Growth of Mycobacteria in Culture

Coinciding with antibiotic searches, Dubos was tackling a different problem. He realized that learning more about mycobacteria would be difficult until these organisms could be grown in a more homogeneous state. Before 1946, standard culture media produced clumps, or pellicles, of bacilli. This made quantitative experiments impossible, because bacteria inside the clumps were exposed to different growth conditions from those on the outside. He and Porter became aware of the shortcomings of this crude culturing technique in Ray Brook's laboratory. Back at Rockefeller, Porter went on to show that the previously accepted multiple shapes of bacilli were really artifacts caused by the culturing technique and Dubos went on to create a new technique to grow the bacilli.[46]

What was needed was a standard bacillus for testing. Despite his criticism of artificial cultures, he accepted them in this case because of the need to study them in isolation. In dealing with a bacteriological problem, a consistent, reproducible bacillus was needed to identify the microbe's components, products, and activities. For the little-understood tubercle bacillus, this knowledge had to be gathered before studying how the bacilli disturbed cells and tissues.

In 1945, the heterogeneity found in growing clumps of bacilli was eliminated by adding detergents to the culture medium. Dubos said the idea came from a popular advertisement for laundry detergents that proclaimed their product, when added to water, caused ducks to sink.[47] Just as likely, the idea came from Hotchkiss who rejoined the laboratory on Dubos' return from Harvard. A supply of detergents was at hand since Hotchkiss was extending his wartime survey of surface active agents, those antiseptics, detergents, and wetting agents that destroy bacterial membranes, among them tyrocidine, a component of tyrothricin.[48] So when asked to suggest something with little toxicity for mycobacteria, Hotchkiss recommended the commercial detergent advertised as sinking ducks and another one called Tween 80.

Davis, Middlebrook, and Dubos found that both Tween 80 and Triton A20 provided rapid growth of mycobacteria by wetting the very waxy cell surface and dispersing the bacilli in culture. Since Tween 80 releases free oleic acid that is toxic to mycobacteria, Davis improved the medium by adding bovine serum albumin, resulting in an oleate-albumin complex that could even stimulate growth of the bacilli.[49]

The new culture media proved exceptional for studying the previously intractable tubercle bacilli. The bacteria grew so rapidly that instead of cultures three to ten *weeks* old containing mixtures of young, old, and even dead bacilli, it was possible to have cultures seven to ten *days* old with bacilli of nearly the same age. These media were used to test the growth requirements of mycobacteria,[50] various environmental effects on growth,[51] and drug sensitivities.[52] They were also used to explore the viability of strains for the BCG vaccine[53] and the surface of the bacterial cells for virulent and immunogenic components.[54] Finally, they were used to establish experimental infection models in mice.[55] The media, however, were not appropriate for clinical diagnosis of disease.

These culture media were a major technical advance that gave highly reproducible results. Some scientists initially criticized using detergents in the media as going against natural conventions for growing bacteria or as creating contaminants that produced artificial bacilli. Dubos acknowledged that these synthetic detergents did not exist in nature, yet he countered that there were many similar natural substances in animal tissues performing the same function, namely, allowing the pathogen to invade and grow.[56] Within a few months, an overwhelming number of scientists easily confirmed the experiments and rapidly adopted the technique. The media are sold commercially as the Dubos-Davis and Dubos-Middlebrook media and remain in constant laboratory use today.

The new technique immediately attracted a great deal of attention. Rockefeller's typically reticent director Gasser told board members in October 1945 that the synthetic media would transform the study of tuberculosis. "On the score of both the speed of growth and the relative homogeneity of the culture," Gasser reported, "the advantages of the new method in the diagnosis and experimental study of the disease are so great as to make the improvement seem nothing less than revolutionary."[57] *Time* magazine heralded the discovery as "the greatest contribution to TB research since Robert Koch first isolated the germ itself in 1882." Dubos was so agitated by this scientific overstatement that he had *Time* retract it a few weeks later.[58]

This discovery, the national news story, and his ensuing lectures on problems that still needed study brought tuberculosis back into the realm of respectable medical research.

Ecological Facets of Virulence

The laboratory studied a broad spectrum of tuberculosis issues, a research agenda that may seem incomprehensible to a researcher today when work on any one aspect can make a career in science. However, when viewed from Dubos' stance on virulence and immunity as the quality of a relationship between pathogen and host, his multifaceted approach becomes more understandable.

Virulence, in his view, is not a discrete property of the bacterial cell. A similar idea was put forward in the 1930s by Theobald Smith, who claimed that a successful parasite is not highly virulent, because if it causes the host to die it then eliminates its reservoir for life. Going beyond Smith's view, Dubos called virulence, and infection itself, a relationship that is entirely changeable. As defined in *The Bacterial Cell*, "Virulence is not a permanent, intrinsic property of a given species. It expresses only the ability of a given strain of the infective agent, in a certain growth phase, to produce a pathological state in a particular host, when introduced into that host under well defined conditions . . . it refers to the disease and to the host-parasite relationship, rather than to some unique attribute of the microorganism."[59]

Specifically, to establish disease a microbe depends on a constellation of independent traits such as communicability, ambient conditions (temperature, acid-base properties, availability of nutrients, gaseous mixtures), toxigenicity, and the ability to multiply. What is virulent for one host may be innocuous for another. Likewise, among the same animal species, and from one individual to another, the host can also exert virulence to foil, thwart, or destroy pathogens. The host's own potential to poison pathogens depends on its age, general physical vigor, and the chemical nature of locally inflamed tissues. Virulence can be related to host or pathogen, in Dubos' view, and can be provoked by many nonspecific external and internal forces or substances.[60]

Contemporary views of virulence are beginning to embrace this iconoclastic view even though, during most of the twentieth century, researchers other than Dubos studied virulence as "a multiplicity of traits that endow the pathogen with its ability to exploit anatomical weaknesses and overcome the immune defenses of the host."[61] More recently, molecular techniques have detected that bacterial genes activated in an infected animal are those responding to stresses (or, in Dubos' thinking, virulence) produced by host tissues.[62] Lederberg has also argued recently that what is needed is a more ecological approach to infectious disease that balances research on pathogenesis, or disease, with equal attention to "salutogenesis," or health. This comment echoes the idea of looking at virulence and immune strategies on both sides of relationships, a notion Dubos began to espouse a half century ago.[63]

Early diagnosis of tuberculosis was another practical problem that Dubos considered "most in need of imaginative study."[64] Until the 1950s, a conclusive diagnosis, even in patients with acute symptoms, took several weeks, since the tubercle bacilli grow so slowly. Once his laboratory had produced a reliable method for growing the organism rapidly, their next step was to look for distinctive traits that indicated whether the bacilli would grow and produce disease. For every trait they detected, they tried to produce a diagnostic test.

The laboratory first pioneered experiments that became standards for defining disease. These included measuring the bacilli's behavior in the size, severity, and duration of lesions as well as the rate and extent of their multiplication in several organs. They established a hierarchy of virulence in human tuberculosis strains, from least to most powerful, and found this order was the same in any infected host, whether it was in guinea pig, mouse, or man. One interesting finding was that virulent bacilli were immediately immobilized by the host's white blood cells, while avirulent bacilli were unaffected.[65]

Other experiments detected a morphological and a chemical trait in strains of bacilli. One, known as the cord factor test, showed that virulent bacilli grow in long strands or serpentine cords, while avirulent strains grow in a disoriented formation of clumps.[66] Laboratory member Hubert Bloch isolated a chemical from the mycobacteria's cell surface that was responsible for the serpentine formation, a substance that more recent investigators have discovered has potent properties to stimulate the immune system.[67] This simple, reliable cord assay remains useful for diagnosing disease, especially when and where modern technologies are lacking. In contrast, the neutral red test, which showed that virulent bacilli adsorbed a red dye from neutral or slightly alkaline solutions, could not completely distinguish between virulent and attenuated strains.[68] Despite their somewhat primitive nature, both diagnostic tests provided important evidence that unique components of the mycobacterial cell wall are important in controlling virulence.[69]

If nothing else, these tests directed the laboratory to search for something that did not exist, a blood test to diagnose tuberculosis. Bolstered by Avery's successes in defining the polysaccharide capsular antigens of pneumococci, Middlebrook and Dubos looked at polysaccharide antigens of mycobacteria. In 1948, they devised a promising blood test, known as the M-D or hemagglutination test.[70] This immunological assay could identify the presence of antibodies to polysaccharide mycobacterial antigens in the serum. The M-D test was applied worldwide with great expectations and was still in prominent use in a large number of scientific and clinical studies as late as 1984. Despite initial optimism and many trials, the M-D test was so difficult to reproduce that it is no longer considered valuable in the diagnosis or prognosis of tuberculosis.[71] Its failure, like antibody tests devised by others, is due to the fact that mycobacteria produce many antigens, both specific and cross reactive, to which the immune system can respond. Even today, no specific antigen has been identified with active disease, and there is still no serodiagnostic test.

Nevertheless, as vital as diagnostic tests are to a physician's armamentarium, none resolve Dubos' basic concern that diagnosing a tuberculous infection is not equivalent to diagnosing disease. Typically, people are unaware that they have been infected and diagnoses are made only after disease has reached an advanced stage. More importantly, detecting a primary, or even latent, infection cannot predict the course of the disease. Will the bacilli disappear and the patient be cured? Or will the bacilli become sequestered and keep their ability to trigger disease at a later time? Or will the bacilli proliferate and produce overt disease? Finding microbes does not foretell their fate. Knowing more about the tubercle bacilli tells more about the formal *cause* of infection but little about the formal *genesis* of the disease. At this stage, Dubos was still searching for a laboratory test to answer these questions.

Immunity to Tuberculosis

Consistent with an ecological stance on virulence, Dubos took a similar nontraditional approach to immunity. He defined immunity as resistance and extended its meaning beyond a traditional focus on antigen-antibody responses and cellular reactions. Just as he believed virulence is not a discrete property of the parasite, resistance is not a discrete property of the host to end or clear an infection. A parasite can also resist or foil the host. Immunity, as interpreted by Dubos, is a variable relationship between pathogen and host in which each has mechanisms to resist the other, to adapt or repel, to win or lose. Or, they can coexist. The outcome—pathogenesis (disease) or salutogenesis (health)—depends on the nature of their encounter.[72]

BCG vaccine. In 1921, French scientists Albert Calmette and Camille Guérin introduced a tuberculosis vaccine, based on principles worked out by Pasteur in 1882, after spending thirteen years weakening or attenuating live cultures of virulent bovine tuberculosis. Named Bacillus of Calmette-Guérin, or BCG, successful tests on nearly a thousand children prompted the League of Nations to recommend the vaccine's use in 1928.[73]

Twenty years later, based on unsettling evidence from his laboratory, Dubos expressed great concern about the safety and efficacy of BCG vaccine.[74] Calmette's original culture was not preserved, and the vaccine was being produced by cultures grown on standard media. Using the new rapid, dispersed culture method for mycobacteria, Frank Fenner and Dubos were able to measure the biological effects of the bacilli in the vaccine for the first time. They identified several substrains of BCG and found they differed in morphology, growth requirements, and immunizing power. The BCG suspensions were not only unstable but contained a mixed population of virulent and avirulent bacilli. This was a clear demonstration that during subcultivation, preparation, and storage of BCG in laboratories around the world, the original culture had undergone progressive variation.[75]

Dubos also worried that live weakened bacilli in BCG continued to undergo mutations, like all bacteria. He saw ominous parallels to antibiotics produced under selective pressure of various culture environments, and repeated a caution first made by Pasteur that "the change from virulent to the avirulent is not always complete and is often reversible."[76] The loss of virulence in creating an attenuated vaccine, he observed, is not due to a bactericidal effect but to environments unsuitable for multiplication. Fortunately, BCG has not been found to be in danger of returning to full virulence.

In further experiments, Werner B. Schaefer, Emanuel Suter, and Cynthia Pierce described a notable variation in the ability of various vaccines to generate an immune response.[77] These studies provided disturbing new evidence that for BCG to be effective, the bacilli in the vaccine must multiply in host tissues and cause a self-limiting disease. In other words, BCG immunizes because it multiplies, not because it produces antibodies. This means the most invasive (or virulent) bacilli are the most effective and the most dangerous. Other scientists and physicians have found that under adverse conditions, the BCG vaccine can produce a mild tuberculosis infection called BCG-osis.

In 1955, Dubos and Robert Debré of the Centre International de l'Enfance in Paris directed an interlaboratory project for UNICEF to establish a scientific protocol to measure and maintain a standardized BCG vaccine. The final report recommended lyophilized, or freeze-dried, cultures of BCG that would contain a fixed number of young, viable cells. The vaccine's efficacy was monitored by using virulence, allergy, and protection tests that were based on Dubos' methods of microscopic enumeration of bacilli in infected organs.[78]

Although a standardized BCG vaccine (using this procedure) was adopted by the World Health Organization (WHO) for use in many countries, the United States did not, and still does not, recommend vaccination with BCG. In 1957, Dubos was a member of the advisory committee on BCG to the Surgeon General of the United States Public Health Service that strongly disapproved of large-scale vaccination due to the decreasing rate of tuberculosis in the country but recommended its use in individual circumstances or in localized outbreaks of the disease.[79]

Killed vaccines. Unconvinced that BCG was safe or had real immunizing power, the Dubos laboratory turned to developing a killed vaccine that owes its activity to a chemical component of the tubercle bacillus to elicit antibodies. Previous experiments suggested this technique might work. Opie had tested heat-killed organisms against *M. tuberculosis* in the 1930s. Dubos had used heat-killed pneumococci and formaldehyde-killed shigellae to prepare vaccines. Just a few years after these trial efforts by Dubos, Jonas Salk had great success using a killed vaccine against polio. The laboratory produced two killed vaccines: one in 1953 used phenol to kill the bacilli and it produced a transient immunity; the other in 1955 used methanol and it elicited about the same level of immunity as BCG. The methanol vaccine actually revived a treatment against tuberculosis that had been tried by two of Calmette's collaborators, Auguste Boquet and Léopold Nègre, in the 1920s. For a time the new version seemed promising, but by 1959, for reasons given below, Dubos lost interest in this project.[80]

Field observations. Walsh McDermott, of Cornell University Medical College across the street from The Rockefeller Institute, was one of the first physicians to evaluate the M-D (Middlebrook-Dubos) hemagglutination tests as well as many new antibiotics being introduced against tuberculosis. In

1953, he invited Dubos to join ongoing studies for treating a tuberculosis epidemic at Many Farms, a remote Navajo community near the Arizona–New Mexico border. Dubos readily agreed despite the arduous travel involved and his precarious health. At the time, air travelers used the single runway airfield near Window Rock. Years later, Dubos reminded John Adair, a cultural anthropologist working at Many Farms, of a particularly trying visit "when Walsh McDermott and I had to push our plane out of the sands of Window Rock before we could get started again."[81]

After several visits to the reservation, Dubos' understanding of the role of environmental and social factors in disease led him to envision the concept of a "hospital without walls." It was this image that inspired the physicians to institute a total medical program on the Navajo reservation to help them become aware of the whole range of community problems, including its cultural values, and to broaden the scope of medical care.[82] In treating tuberculosis, the Navajo medicine men accepted and were impressed with how isoniazid healed the disease, and the physicians accepted the traditional Wind Chant ceremonies as relevant to the healing process. This trust between tribal medicine men and modern physicians, McDermott said, gave him "a deep sense of the fact that concepts of illness are more closely integrated with spiritual concepts." Alluding to Dubos' views, he added that "with respect to environmental and emotional influences on what happens to the tuberculous patient, the concepts of Navajo singers and of our own people are really not too far apart."[83]

McDermott and Dubos contemplated, but then rejected as unfeasible, using the methanol-killed vaccine to reduce the rampant spread of tuberculosis in this native population. They actually established a pilot plant at Cornell to purify and produce large quantities of this vaccine. An extremely hazardous process, it involved growing fifty liters of tubercle bacilli every day, and the fine spray thrown into the room by centrifuging the bacilli resulted in violent allergic reactions among the technicians. The project was quickly aborted.[84]

Dubos' ecological thinking on vaccines, in general, stands as a cautionary tale in the context of current practice. He believed a vaccine should not be used until mechanisms of immunity were known and methods available to measure a true protective immune response: a positive tuberculin test does not necessarily measure active disease—only allergy or previous sensitization to the bacilli. Moreover, he believed a vaccine's efficacy should be evaluated in different populations living under different social conditions, and he predicted that responses to the same vaccine would vary greatly depending on whether it was tested in a rural population with a low incidence of disease, an industrial population with low standards of living and high exposure, or a population coming into contact with the disease for the first time.[85] Still, in 1995, according to tuberculosis researcher Barry Bloom, the efficacy of the BCG vaccine remains unknown and the "absence of any confirmed and consistent measurable laboratory correlate of protection in humans remains a major challenge to research on mycobacterial vaccines."[86] While the newer cell-free vaccines eliminate safety issues caused by live organisms, their ability to give long-lasting protection remains unknown.[87] More germane, in Dubos' view, people are as different in their susceptibility to tuberculosis as they are in their response to the BCG vaccine.

In 1994, WHO declared tuberculosis "a global emergency," the leading killer worldwide today, causing eight million cases and two million deaths each year. The new rise in the incidence of tuberculosis was due less to biological reasons than to perilous social factors: homelessness, substance abuse, the deadly co-infection of HIV/AIDS and *M. tuberculosis* that enhances viral replication, and strains of tubercle bacilli that are resistant to multiple antibiotics. The shift from treating tuberculosis patients in sanatoria to allowing patients to self-administer drugs at home has led to noncompliance, treatment failure, and the emergence of resistant strains. In addition, some observers are now suggesting that impaired public health systems in many regions may be incapable of containing or suppressing a new White Plague.[88]

For not so surprising reasons, the tubercle bacillus remains a master teacher. The scientific complexities of tuberculosis that burden medical research today reveal the prescient observations of Dubos. There was such complacency about the disease being under control in the 1960s that scientific research essentially ended. Among the six needs identified recently by WHO to confront this global emergency, two are for controls that were firmly in place before the 1950s and then abandoned or neglected: quarantined patients and active public health organizations to monitor and respond to outbreaks. Two other problems result from complacent medical and public health behavior: drug resistance and directly observed therapy of unquarantined patients. The remaining two needs are for methods to detect and to prevent tuberculosis. Echoing Dubos fifty years ago, WHO effectively acknowledges that neither a diagnostic test nor an effective vaccine exists.

What remains vital about Dubos' laboratory experiments on the tubercle bacillus is that he reached a deeper level of unknowns in microbial disease: virulence and immunity are relative. Responses to the pathogen, infection, antibiotic therapy, or a vaccine, he advised, depend on an individual's social and environmental habitats, general health, nutritional status, and emotional well-being.[89]

By 1951, Dubos decided that learning more about the biology of the tubercle bacillus was unproductive. As he wrote Jacques Monod, he was becoming "preoccupied with problems of general physiology and pathology as they relate to phenomena of infection. I am not up to date with facts of microbial morphology and physiology, and I'm afraid to repeat myself, especially the old mistakes."[90] Here was a rare admission that he had spent too much effort on the pathogen and that these experiments had made him aware of a need to understand two types of events, those that allow infection to take place and those that turn infection into disease. Significantly, at this point, he was beginning to champion the theory that a microbe is a prerequisite but not necessarily a determinant of disease.

4 Mirage of Health: Infection versus Disease

To regard any form of life merely as slave or foe will one day be considered poor philosophy, for all living things constitute an integral part of the cosmic order.

RENÉ DUBOS

AS RESEARCH ON TUBERCULOSIS PROGRESSED, Dubos wondered how to broaden his approach to the disease. Outwardly, the laboratory was making important advances. Inwardly, his characteristic restlessness and skepticism made him reconsider this course. The catalysts for change came from outside the laboratory, and none provided a more ideal atmosphere for this transition than his once-abandoned farm.

Planting Trees, Planting Ideas

Long Sunday walks in the country were one of Dubos' few worldly pleasures. On returning from Harvard in 1944, his future projects and personal life were unsettled. In a letter to his mother, he confessed that the public success of his earlier studies gave him "a popularity that I find rather disagreeable. . . . After having tasted a little of the world, I have decided to find again the most absolute calm and I have very much restricted my circle of friends." With wartime gasoline rationing, it was pleasant to live just across the street from the Institute and to walk among its sycamores, flower gardens, and lawns that were unique to New York, "like an oasis where life remained gentle enough." But not simple or quiet enough.[1]

On Sundays, after working a full six days, he traveled by train an hour or more into the Hudson River valley. Starting at a different town each week, he sauntered without maps or destinations along mountain roads that were wild and nearly deserted, returning at dusk to the city. He enjoyed society in town, but out of doors nature was company enough. The open spaces allowed him to feel, think, and be himself, free of impediments, inconveniences, and other people. He was content to read nature carefully, laying in a stock of ideas while searching for "an ideal place in the country." There was such pleasure in looking, he also told his mother, that the searches were more enjoyable than actually owning something. "I can be happy with nothing provided I have peace."

These yearnings for simplicity intensified when, at the age of forty-five, he married Letha Jean Porter. She became a perfect companion on his Sunday walks for she loved the countryside and wild nature more, and social life less, than he did.

Jean was born in Upper Sandusky, Ohio, on 1 January 1918. Her childhood was also one of delicate health that included a mild bout of tuberculosis. Both of her parents were educators with masters' degrees in English. Her father, Elmer Porter, spent a decade as superintendent of schools in Columbus, Ohio, before his early death in 1937. Nora Wills Porter, her mother, supposed Jean would someday follow her as dean of women at a small college, but she preferred instead the tactile nature of experiments in a laboratory and graduated from Ohio State University with a degree in biology. As a budding microbiologist with four years' experience in a pharmaceutical company, she sent a telegram to The Rockefeller Institute, pluckily announcing her impending arrival on 11 April 1942 and requesting an interview for a job as a laboratory assistant. During the train ride from Columbus, she read about Dubos' discovery of antibiotics and was startled to learn he would be the one to interview her for an opening in his laboratory. Her acceptance and arrival in his laboratory took place during the few tragic weeks surrounding Marie Louise's death.

For four years, Jean served as personal laboratory assistant to Dubos, accompanying him to Harvard in 1942, where she was a scrupulous reader of *The Bacterial Cell*, and back to New York in 1944. He described Jean to his mother as "blond, with blue eyes, rather tall, incredibly thin, nearly a girl" at twenty-eight years of age. Jean "is very shy, very reserved, never speaking in public, and always with a low voice. Her gaiety is all inside and her temperament most agreeable."[2] Independent and creative, Jean was an admirable colleague, taking part in all his research and writing. He credited her with balancing his visionary and sometimes romantic goals for the future with more visceral and spiritual values. Her inclinations and wisdom complemented his, and with her calming personality, she always brought out something new.

Jean and René were married on 16 October 1946 at The Church of the Epiphany, a few blocks north of the Institute. This intimate ceremony was witnessed only by Oswald Avery, Cynthia Pierce, and Jean's mother. Once married, Jean left the laboratory, abiding by the Institute's rules that did not allow wives to work with husbands.

Together René and Jean searched for a small property with the appealing aspects of woodland and farming that both had known in their childhood years. Finally, during a walk along the Appalachian Trail near Garrison, New York, they found and bought an abandoned farm. This was neither a rural nor a pastoral landscape. Its ninety acres extended along rocky slopes of Denning Hill in the Hudson Highlands. The narrow valley is traversed by the Old Albany Post Road, where milestone sixty-four dated 1680 marks their property. The farm's fields and orchards, bounded by stone fences, once produced crops, including watermelons, that were shipped by boat downriver to New York. There was nothing sensational in this wild retreat; it was a gesture to return to plain living, a place to gain strength from the restless, nervous, bustling, and unimportant details of life in the city. Leading as simple a life as possible, they regarded it not an escape from reality but the embodiment of it.

The adventure of the new landowners began with a realization that they could not take the charming countryside for granted. During the long postwar deprivation of materials that precluded building a house right away, René envisioned cultivating this landscape with individual trees that would stand out from the surrounding wild forest. As he wrote his mother, "if we live long enough, we would have a large beautiful park, with natural features, but not a virgin forest. . . . There is nothing that gives us more pleasure than to work, dressed like peasants in rags and doing as we please, to conserve its wild, natural aspects."[3] Jean noted that René was "realizing his submerged desire to be a forest ranger and landscape architect. He has done a tremendous amount of clearing and has great dreams of opening vistas and setting out fine trees."[4] The first trees he planted were the little spruces that he had "carried back to New York in a brown paper bag" after a visit with Marie Louise in the Adirondacks. Long after these trees had grown large and healthy, he told the friend who helped find them, "I measure the length of my life against them and they help me to keep the past alive."[5]

Dubos thrived on the physical labor of scouring the fields and forest floor to find seedlings with promise and transplanting trees smaller than weeds into just the right environment of soil, sun, and moisture. It took "ferocious energy" prying countless boulders from the soil to make room for tree roots, yet he loved digging stones for their own sake. For trees that began to grow, he kept the soil loosened. For those that grew even more, he made tepees from trimmed branches to ward off animals and to keep out winter's cold and summer's sun. Paths were cleared to get to these trees during visits that mimicked a physician's rounds. Tracing their lineage and monitoring their vital signs, he took pains to note every change in each developing tree. When late summer streams dried up, a frequent occurrence, he hauled buckets of water to them. Those trees that grew quickly pleased him most, but he was patient with the others, except when spring came. Then, "I have an eagerness to see how much my little trees have grown, which makes me very eager to be there next spring, to see whether they have grown a little more."[6]

Once a tree was established, he removed its fencing and turned his attention to its environment. Pruning shears in several sizes were wielded to open up the understory, providing air, light, and space for the tree to grow. The idea was to create a hospitable terrain while shaping the trees and exposing pleasantly shaped rocks that had been hidden from view. He used the trees to sculpt his visions of a "garrisonian landscape." This was accomplished not so much by removing undesirable vegetation or planting orderly gardens as by assembling trees and plants to create contrasts of shade and sun that he admired in the sunlit forests of the Hudson River School artists. He experimented with muted greens, a favorite color, in places where breaking mists on fields met laurels and junipers at the forest's edge.

As time went on, some trees grew so hardily and pleased him so much that he named them. "Proust" was a freestanding elm that provided cool, dark shade, and its straight trunk supported the couple while Jean perfected her French by reading aloud *À la recherche du temps perdu*. These trees offered nurturing places that invigorated his writing and dreaming. Knowing he would not live long enough to see his trees reach maturity, he enjoyed them while they grew and had faith they would be enjoyed by others. Planting trees, he believed, is an "act of faith in the collective future of man."

A New Rockefeller Environment and New Experimental Medicine

Also in 1946, catalyzed by challenges from the Institute's director, physiologist and Nobel laureate Herbert Gasser, Dubos began to think about a different approach to science. Gasser's leadership bridged the economic depression of the late 1930s, World War II, and the unsettling changes in the funding of scientific research at the end of the war. His medical training and grasp of mathematical and physical sciences equipped him to lead the Institute's transition to study human biology, not just disease. Gasser wanted the Institute "to intensify investigation of fundamental life processes at the level of the cell and its constituents." He committed to a goal of producing "scientific capital," whose exploitation could be left to the "multitude of workers." One of his decisive principles was that there were two times for working on a problem—before anyone has thought of it and after everyone else has left it. Within this small community, members were being asked to perform research that was "path breaking" as opposed to "path following."[7]

Dubos took this challenge to consider an approach that would move beyond the germ theory and embrace the concept of plasticity that he presented in *The Bacterial Cell.* Bacterial studies were shifting rapidly toward the analytic breakdown of component parts—from studies on the bacterium as a whole to components of the cell (cellular biology, bacterial genetics), and then to individual molecules (molecular biology). Secure in an institutional environment that provided freedom to explore something that fit his own curiosity and motivation, he took Gasser's challenge to confront a fundamental problem of biology: bacterial interactions in health and disease.

During the same year, Gasser and Peyton Rous appointed Dubos to the editorial board of *The Journal of Experimental Medicine.* Since its founding in 1896, the journal had reflected the views of its original editors, William Welch, Simon Flexner, and Rous, all physicians and pathologists. The addition of Dubos was a significant departure, and he cautioned Rous that while he felt competent in bacteriology, virology, immunology, and some aspects of biochemistry, he was "grossly ignorant of pathology and physiology." Rous replied that he was appointed not only for his scientific qualifications but for his views about the Institute's uniqueness and success, particularly his often expressed view during lunch room debates about whether medicine was a science or an art. That was an irrelevant question, Dubos had argued, because the Institute's scientific staff, including the physicians, were simply satisfying their "hunger for facts." Rous admired this phrase for capturing the community's predominant spirit that he believed "made it an organism, not an establishment."[8]

For the next twenty-five years, Dubos engaged in weekly reviews of papers that exposed him to various approaches to experimental medicine, an assignment that, over time, influenced Dubos' research, the nature of the journal, and indirectly, the evolution of the Institute. Initially, he found himself balancing Rous' traditional focus on pathology with Gasser's focus on physiology to expand medical science into more basic life processes. Within two years, Dubos was exerting his ideas about the kind of medical research that was broad and novel enough for the journal. In correspondence with Rous, he rejected papers for which he could not "recognize the intellectual basis" or that had "no fundamental, general significance." He declined papers of a specialized nature, for example, those on biochemical aspects of individual bacteria. However, he accepted other highly specialized papers that "deal with a general problem—the nature of viruses—which the editors of *The Journal of Experimental Medicine* have long regarded as one of broad interest and fundamental nature." He turned away "all the good papers on chemotherapy," since the field was becoming too popular, "in favor of those contributions that represent a new departure, or offer promise of great practical usefulness."[9]

A more focused challenge came from physician-in-chief Thomas Rivers, who helped Dubos think about pathology and physiology of many infectious diseases. In 1946, Rivers, who was preparing a reference book, *Viral and Rickettsial Infections of Man*, asked Dubos to prepare a companion volume, *Bacterial and Mycotic Infections of Man*. Both texts were very successful, going through four editions well into the 1960s. Their novel aim was to prepare medical students who would become physicians or medical researchers rather than to educate microbiologists. Dubos showed no fondness for the task of planning and editing a volume authored by nearly three dozen specialists. He reluctantly acted as editor in this case because he was eager to reach medical students with his ecological outlook, stated in the preface, that medical microbiology is "the study of host-parasite relationships and not that of microorganisms alone."

By 1948, when the first edition appeared, he was already boldly stating that the ability of an organism to cause disease depends less on the biochemical and biological characteristics of bacteria or fungi than on the physiological state of the person who encounters microbes. This argument reveals that Avery's considerable influence on him was dwindling. That same year in his O.T. Avery Lecture, Dubos urged a study of infection from the "point of view of the ecologist" and summoned a path, suggested by Theobald Smith, to examine the "delicate equilibrium between invader and invaded host." The goal was to study "those structures and reactions at the level of which the animal and bacterial economy influence each other."[10]

These views were not shared completely by the premier bacteriologists and infectious disease specialists who wrote individual chapters for the book. These scientists had vested interests in classical techniques of bacteriology to make specific diagnoses and to produce drugs or vaccines. Complicating the matter further, Dubos had shifted his thinking on infection and disease, but his research strategies were still focused on the tubercle bacillus. There was, he realized, "a very small body of experimental evidence, hardly any techniques of investigation, and certainly . . . no immediate practical application." With little to guide him and no experiments in mind, how should he pursue such an approach?[11]

In the Shadow of Pasteur

His approach, at first, seems strange. He decided to write a biography of Louis Pasteur, a decision that changed his scientific life. This was curious because Pasteur had not studied tuberculosis and, in the fifty years since his death, had been the subject of exhaustive biographies. Yet Dubos' carefully reasoned analysis of the conflicts in Pasteur's life shows why he turned to this early master of bacteriology and, more relevantly, to the time before views of a biochemical unity of life and germ theory of disease were formulated. As he once advised, it is always profitable to read past masters because "their writings often suggest avenues of thought not yet explored."[12] *Louis Pasteur, Free Lance of Science* is an eloquent story of a scientific life and gives penetrating insights into Dubos' own future studies of infection and disease.

Pasteur and Dubos were similar in many ways. Both were born and raised in French villages and were plagued by ill health during their careers. Both began with biochemical problems and shifted to larger issues of infectious diseases. In the laboratory, both were quiet and contemplative, devoted to their work, and acted almost exclusively alone. When making public appearances, both had a gift for reaching large audiences with picturesque and vigorous arguments that were expressed with typical Gallic charm and persuasion. These affinities were sufficient to attract Dubos as an empathetic, although critical, biographer.

There was also a profound difference between these two scientists in their scientific environments and attitudes toward their achievements. Pasteur focused on those aspects of science likely to produce concrete results, even though he pondered problems that he neither pursued nor considered ripe for solution. Pasteur relished success that came from his dramatic controversial public debates and scientific demonstrations. Dubos reproached Pasteur for not knowing "how to conceal the great pride he took in his discoveries. . . . For him, social recognition was the symbol that he had fulfilled his calling."[13] This comment suggests an influence of the Rockefeller culture on Dubos, under Flexner and especially Gasser, that in contrast valued definitions and identification of scientific problems over practical applications, the kinds of problems that are often neither understood by contemporaries nor acquire import until much later. Unlike Pasteur, Dubos was apolitical and self-effacing; for example, in his accounts of the discovery of antibiotics he always emphasized penicillin and rarely mentioned gramicidin. Dubos debated scientific controversies only with colleagues in the lunch room or laboratory; he neither courted scientific recognition nor outwardly relished the public attention that his discoveries brought.

The Pasteur biography was written quickly during an anguished 1948 while his wife Jean recovered from a mild case of tuberculosis at a Pomona, New Jersey, sanatorium. For the most part, it was written under an apple tree in the orchard where their house was taking shape. With a long open view of the valley and pond below, he was, he told his mother, "in the midst of greenness and silence I have dreamed of all my life. I am profiting from it in writing my book on Pasteur which is now well on its way."[14]

The biography's actual provenance can no longer be traced, although at least three sources can be ruled out. It is unlikely that a publisher asked for this biography, since Dubos said several publishers rejected it, one, he said, because he had not described the shape of Madame Pasteur's legs. Furthermore, Pasteur was not a childhood hero for Dubos. Indeed, during his entire life, while he spoke admiringly of Winogradsky, he did not attribute a concrete influence of Pasteur on his scientific career and his few references to Pasteur concerned mostly the fermentation experiments. Also, the book's origin can only remotely be traced to a question from a Harvard colleague, biologist and chemist Jeffries Wyman, about why Pasteur was so famous when he "never did discover any of the fundamental laws for which the nineteenth century is so well known." This remark, however, may have prompted the biography's organization by the topics of Pasteur's experiments, from crystallography and fermentation to spontaneous generation and vaccination.[15]

Many extrinsic aspects of this biography stand out as uncharacteristic of Dubos. He had not previously written in such a narrative style. Even his two later biographies were distinctly different: a second one of Pasteur in 1960 for high school students and a scientific profile of Oswald Avery in 1976. Dubos' lifelong method of writing involved giving lectures to test themes and ideas and eventually reworking them into book chapters. In this case, the manuscript emerged in six months as a uniform whole. Moreover, the narrative contains peculiar words and idiomatic phrases that Dubos almost never used in his classic French manner of speaking or writing English. Its overly fluent sentences are more refined and limpid than he used in *The Bacterial Cell* just a few years earlier and in public lectures published during the same time. Such polished writing may of course be attributed to a skillful editor. A more reasonable speculation is that Dubos, consciously or unconsciously, synthesized passages from Pasteur's writings or other biographies and may have allowed the veneer of their storyline to dominate while he extracted more theoretical implications beneath Pasteur's scientific evolution.

Dubos in fact proceeds to recount Pasteur's compelling life story while grappling with the nature of experimental medicine. This theme was inspired by a debate he had read in 1944 concerning experiment versus experience as techniques for acquiring knowledge. The debate took place in 1882 between Pasteur, on entering the French Academy, and linguist, philosopher, and historian Ernest Renan.[16] Dubos originally planned a book exploring Claude Bernard's distinction between a man of research (experiment) and a man of science (experience and knowledge) as it played out in this debate, a topic he only broached in *The Bacterial Cell.* When the book became a biography of Pasteur instead, the debate forms its conclusion and ties together the threads of Pasteur's research into Dubos' vision of what lay "beyond experimental science."

This debate captures what Dubos valued in Pasteur and even more highly in science, namely, to understand the nature of life and to give more meaning to human existence. Pasteur, like physiologist Bernard, defended the supreme, though limited, role of experimental science while acknowledging that many mysteries of life reside outside physicochemical phenomena. Renan countered that human feelings, experience, and even religious dogma are amenable to scientific analysis by using what he called the "little conjectural sciences" of sociology and history. Then echoing Bernard, Renan claimed "the great experimental *principle* is doubt, not a sterile skepticism but rather a philosophical doubt, which leaves freedom and initiative to the mind."[17]

The real power of this biography derives from the questioning criticism that Dubos applied to this tension between experiment and experience in Pasteur's work, an attitude that keeps the book from idolizing the scientific triumphs. His skepticism of Pasteur can be traced back as far as his student years in France. Dubos recalled how science teachers and textbooks portrayed Pasteur as a benefactor of mankind with his discoveries of great practical importance, and literature texts portrayed Pasteur and other scientists as having a lofty place in civilization due to their power to solve the enigmas of the world and the real nature of life. With his interest in history at the time, Dubos confessed these ideas left him "lukewarm" about "the utilitarian aspects of science" and confused about the relative importance of literature and science. What heightened this skepticism was reading Renan's preface to *L'avenir de la science*, written in 1848 when he was twenty-five years old, but published only in 1890. In the intervening years Renan had tempered his original overconfidence in science and concluded that "the main contribution of science will be to deliver us from superstitions rather than to reveal the ultimate truth." The influences of these early cultural experiences on Dubos, he acknowledged, made it "extremely difficult, if not impossible, to define science, to delineate the traits that differentiate it from other human activities."[18]

This skepticism about Pasteur as a hero was further piqued in 1935 during a lecture at The Rockefeller Institute by Pasteur's grandson, Pasteur Vallery-Radot, and during a lengthy visit by him to Dubos' laboratory in 1944.[19] Both Vallery-Radot's lecture, "The Logical Sequence of Pasteur's Work," and the view that his grandfather could do nothing wrong were exposed by Dubos as false. He instead depicted Pasteur as taking an erratic course determined more by fruits of intuitive visions that came long before experimental evidence could verify them.

The dual meaning of *free lance* in the biography's subtitle conveys precisely this tension and allows an evaluation of Pasteur's unique traits that were outside the realm of hero worship or public acclaim. In a negative sense, Dubos portrayed Pasteur as a free lance in its root definition of a mercenary who offers his services with no fixed allegiance. Reflecting the nineteenth-century scientific environment, Pasteur had sacrificed great theoretical issues to pursue such industrial problems as those pertaining to beer, wine, and milk. He became, in Dubos' harsh judgment, "the champion of a cause rather than an intellectual giant," and "the prisoner, almost the slave, of limited and practical tasks." In a positive sense of free lance, even though "Pasteur was made to behave as a bourgeois and to accept the code of experimental science . . . he was by temperament an adventurer."[20] Pasteur was thus praised for perceiving far-reaching implications of isolated experimental facts and translating his practical achievements into general laws of nature. In Pasteur's hands, Dubos wrote, "the experimental method was not a set of recipes" but a living philosophy adopted to ever-changing circumstances.[21] This connotation of free lance also offers an intriguing self-portrait of Dubos.

Throughout the biography, Dubos reveals concepts that Pasteur expressed but never pursued, particularly those related to environmental influences on infection and disease. These were experiments Pasteur "could have" done and the trends he "could have" followed, all of which were abandoned during his restless march forward. Significantly, these were concepts that Dubos went on to test in his laboratory:

- After demonstrating silkworm diseases were caused by microbes and designing successful methods to protect the worms from infection, Pasteur suggested that in the future "I would direct my efforts to the environmental conditions that increase their vigor and resistance." Based on a single dramatic test of temperature on susceptibility, he showed the presence of a pathogen was not necessarily synonymous with disease.[22]
- Instead of developing a vaccine for anthrax in sheep, Dubos commented Pasteur could have focused on the nature of infection rather than its causative organism and this "might have yielded results which now remain for coming generations to harvest."[23]
- Pasteur made "prophetic observations" on how a latent or carrier state of infection might determine outbreaks and destructive epidemics but did not pursue how one animal species could serve as a reservoir of infection for another species.[24]

- Pasteur observed a loss of virulence in several pathogens that allowed him to create vaccines based on attenuated cultures. His further observation that the reverse could have happened, generating "a new virulence and new contagions," suggested to Dubos that changes in virulence might be responsible for the genesis, evolution, and cycling of epidemics.[25]
- Pasteur assumed that chickens afflicted during a cholera epidemic were killed by multiplying microbes. He suggested, but did not test, whether a toxin or "narcotic" might be producing their death. Such an experiment, Dubos commented, "could have heralded a new phase in Pasteur's scientific life."[26]
- Dubos dared to specify some discoveries Pasteur might have easily made. Amusingly, and without self-reference, these are areas where Dubos himself had made such discoveries, specifically, in soil fertility, the therapeutic value of saprophytic organisms or "bacteriotherapy," and antitoxic immunity.[27]

The Pasteur biography was a commercial success and assured Dubos' popularity as an author of many other books for the general public. Six thousand copies were sold within the first ten weeks, surprising the editor, who felt the book was limited to a scientific audience. On its publication in 1950, *The New York Times* called it "one of the best biographies of the year."[28]

Inside the Institute, moreover, Dubos was encouraged to look beyond experimental evidence to learn how Pasteur visualized microbial life as integral to processes in health and disease. The germ theory, he now boldly concluded in the biography, "is not a philosophical theory of life, but merely a body of factual observations which allows a series of practical operations."[29] Only when republishing the biography in 1976 did he describe Pasteur as an inadvertent ecologist whose wisdom rested on understanding how "all forms of life are integrated components of a global ecological system."[30] What was not so apparent in 1948, however, is how writing this biography turned Dubos into an advertent ecologist who from this time forward was invigorated to embark on some roads that were seen but not taken by Pasteur. What was apparent in 1948, echoing Pasteur, was that remedies treat established *disease*, but this was not the most useful solution for controlling *infection*. As Dubos would argue for the rest of his life, any improvement in health in the future, as in the past, will come from preventing conditions that lead to infection rather than from interrupting the mechanisms of disease after it occurs.

Microenvironments of Inflammation

In the words of an old physician, "Everyone has, has had, or will have a little tuberculosis" at some time during his life. It was well known by pathologists doing autopsies and by physicians taking tuberculin tests that very few people escape infection with such a ubiquitous microbe as the tubercle bacillus. The real danger is that most people are completely unaware of their infection. When an individual becomes infected, the tubercle bacilli may be killed rapidly by the local tissues and/or macrophages; or the bacilli may grow within a few weeks into lesions that produce active disease, which may either disappear or turn fatal; or the bacilli may remain alive yet dormant inside macrophages, where they remain poised to trigger active disease at any time. Individuals with latent infection show no clinical symptoms of illness and appear to be in good health, just as Marie Louise Dubos did for nearly forty years. Nearly half of all cases of tuberculosis arise from reactivated silent infections.

Within a short time, Dubos enlarged his sphere of research to focus on interactions between the tubercle bacilli and its hosts. This phenomenon had been studied since Pasteur, but the aspect of host-parasite studies that Dubos embarked on was then considered "a dark chapter in physiology," according to Arnold Rich in *The Pathogenesis of Tuberculosis* (1944). "Aside from the question of a specific curative agent," Rich noted, "the most important problem in tuberculosis today appears to be the problem of individual native resistance. Why do only *some* of the enormous number of individuals develop progressive disease?" Methods were needed to determine the "nature of the tissue environment" and factors that govern the growth of the tubercle bacilli in the body.[31]

Dubos took the approach that "the bacillus, although a necessary condition to clinical disease, is not in itself sufficient. *The microbe needs a fertile soil.* It produces clinical tuberculosis only when the individual's hereditary constitution, physiological disturbances, emotional upsets, overwork and other excesses prepare the ground" (emphasis added).[32]

A different test system was needed to look at the fertile soil of tissue environments in the body that enable the bacillus to take root. Dubos made a controversial decision to use mice exclusively as the experimental host for tuberculosis. Early investigators, including Koch, who was the first to try the mouse, were interested in progression of pulmonary disease, so they preferred guinea pigs and rabbits, which are so acutely susceptible to experimental infections. Nevertheless, controversies continued about which animal models most closely approximated human disease. Dubos was never under the illusion that the mouse reproduced true clinical tuberculosis, just as the highly artificial pneumococcal infections in mice and monkeys in Avery's laboratory were never entirely analogous to human lobar pneumonia.[33] To scientists who opposed the mouse model, including Rockefeller's Eugene Opie, he argued that animal experiments were highly simplified systems and not meant to duplicate faithfully what occurs in nature. The mouse was a "reagent," or in a recently coined phrase a "fuzzy test tube," designed to analyze certain aspects of a complex problem.[34]

Not only did the laboratory get good results using mice for tuberculosis experiments, but mice were smaller, more convenient, easily tested in large numbers, and much less expensive than larger animals. Another benefit was the newly available mouse strains of known genetic makeup, which reduced variability among experiments and permitted comparative studies of susceptibility among strains. The strain they found with the highest susceptibility, the C57 Black mouse, is still the model of choice in tuberculosis studies today.[35]

Above all, the mouse model provided an opportunity for the laboratory to study a phenomenon Dubos labeled "microenvironments of inflammation." Their working hypothesis, once again reflecting Winogradsky's philosophy, was that the biochemical compositions of tissues are as individual, and their activities as intricate, as those found in various soils. In addition, reasoning ecologically, any disturbance to the tissues would change the available nutrients and this milieu would in turn change the microbes and their behavior.[36]

The experimenters looked at the biochemical nature of the locales or microenvironments where a variety of circulating microbes gather, the fixed tissue cells react to microbes, and substances of cellular origin concentrate to control the inflammation. For example, it was known that lactic acid is produced during normal cell metabolism and is generally toxic to bacteria, but the laboratory unexpectedly found that during inflammation lactic acid accumulates in the local tissues and takes on the role of a normal host defense.[37] In addition, they found that some chemicals inhibited the growth of tubercle bacilli while others promoted their growth. Still other environmental conditions, such as decreased oxygen, increased acidity, and accumulating metabolites, were found that could resist an infection by creating necrotic areas unsuitable for a pathogen's growth.[38]

Other experiments looked at how antibiotics interact in these microenvironments, a situation that becomes critical when tissues harbor tubercle bacilli, or even staphylococci, in a latent form. It was found, for example, that some antibiotics, such as streptomycin, can kill bacilli outside cells but cannot penetrate phagocytic white blood cells to abolish sequestered bacilli. Whereas many antibiotics work only when microbes are proliferating during acute phases of disease, isoniazid can kill active but not resting intracellular bacilli. Furthermore, slightly acidic tuberculous lesions as well as some natural tissue substances prevent antibiotics from working. In humans or animals receiving antibiotics, Dubos observed, the dormant bacteria, or persisters, "remain fully susceptible to the drug in use," yet they are not touched by it. The danger from viable bacilli that escape sterilization and persist within closed lesions is that they can be reactivated by any number of elements when microenvironments change to favor their growth.[39]

Fourteen years after his initial warning of antibiotic resistance, Dubos lectured on microenvironments to a medical community alarmed by the near epidemics of resistant strains of staphylococci in hospitalized patients. The "organisms that initiate the disease," he said, "may become profoundly altered in . . . an abscess [and] may have lost temporarily some of the very properties that had first endowed [them] with invasive power."[40] He also warned agricultural scientists of dire consequences from feeding antibiotics to farm animals, especially when they are kept under unsanitary conditions or given deficient diets.[41] This warning should not have been so surprising since at the time it was known that antibiotics fed to chickens turned up in eggs and penicillin given to cows appeared in pasteurized milk, thus entering the food chain and exposing humans to low levels of these agents.[42] No matter whether the drugs were given as prophylactics or as "growth factors," he presented the greater worry that this practice would lead to earlier appearances of antibiotic resistance and breed completely new diseases by altering "the very characteristics" of the microbes themselves.

At the same time, Howard Florey and others were blaming antibiotic resistance on careless hospital workers and cross-infection.[43] They urged the creation of other antibiotics with broader striking power to fight the resistant strains. Dubos, however, condemned such broad-spectrum antibiotics as "gun-shot [*sic*] treatments" and predicted they would create other diseases while potentially curing ones for which they were intended. "Broad-spectrum antibiotics," he said, "can become the cause of man-made diseases by permitting the multiplication of certain microbial agents which cannot compete" with the microbiota in normal tissues.[44] Unfortunately today, as he predicted, the too-efficient curb on microbes from using antibiotics is spawning disastrous secondary effects on both animal and human populations.[45]

Ecology of Infection versus Disease

The studies of microenvironments led to a closer look at the problem of equilibrium in latent infection. In 1934, Theobald Smith had hypothesized that the human species had survived for countless generations in association with tubercle bacilli as a ubiquitous member of the environment. In general, the parasite is efficient and the host is tolerant, so if this "efficient parasite" had not achieved an equilibrium, Smith argued, the human race would have disappeared long ago. A substantial amount of historical experimental evidence, gathered by Dubos' first graduate student, physician Harold J. Simon, showed that latency or attenuated infection is a universal occurrence and "only rarely results in infectious disease."[46] For Dubos, understanding which factors turn a quiescent infection, or benign parasitism, into fulminating disease applies equally to every microbe and disease, not exclusively to tubercle bacilli and tuberculosis: "Only when something happens which upsets the equilibrium between host and parasite does infection evolve into disease. In other words, infection is in many cases the normal state; it is only disease which is abnormal."[47]

The laboratory turned to exploring *general* stresses in the host's external environment that might turn normally peaceful agents of infection into potentially fatal ones. This work recalls Dubos' 1934 experiments in Avery's laboratory in which detrimental environmental factors lowered the physiological status of monkeys and made them susceptible to lobar pneumonia. Its rationale also reflects experiments by Dubos at Harvard in 1944 with a group of physicians who studied gas gangrene and toxic shock. Alarmingly, they found that common intestinal bacteria, which are normally destroyed on entering the blood stream, become lethal following severe trauma, wounds, or radiation exposure.[48]

The new experiments further amplified the importance of microenvironments, showing that when an otherwise normal but infected individual is stressed, the metabolism of tissue cells becomes abnormal and thus incapable of producing an adequate biochemical defense.[49] The scientists challenged animals with nonspecific factors such as trauma, fatigue, simultaneous illnesses, metabolic disorders, nutrition, or toxins, all of which represent common accidents or stresses of human life. Although Dubos recognized that these stresses work through any number of indirect, biochemical channels within the body, which were not investigated, the experimental focus was kept on the break in tension between infection and disease. As in previous experiments, the laboratory measured disease in survival time and in bacterial counts from organs at various intervals following the stress.

The studies described in the following pages produced some of the most important and provocative results of Dubos' scientific career. In many particulars, they form the basis of his expertise and later warnings related to environmental determinants of human disease.

Concomitant infections. Dubos tested a hypothesis proposed by Rockefeller colleague Richard Shope that many deaths during the 1918 influenza epidemic occurred because the flu virus served to reactivate a tuberculous infection. The scientists took two groups of mice infected with tubercle bacilli and exposed one to a flu virus. Both groups developed tuberculosis but the group exposed to the flu virus became ill more rapidly and extensively.[50] In the case of superimposed infections, where one agent alone was not severe enough to cause serious disease, the two together overwhelmed the host's responses and turned latent infection into active disease. These results added a clue to Dubos' view of the cyclic nature of epidemics: while many people are naturally infected with a variety of organisms, very few individuals develop clinical disease until they are stressed by other factors, including other microbes.

Metabolic disorders. Individuals suffering from diabetes or from starvation were known to be more susceptible to tuberculosis. Dubos suspected this occurred because these disorders produced slightly acidic conditions in the tissues, an observation confirmed by in vitro tests where certain organic acids facilitated the growth of tubercle bacilli and staphylococci. Tests with mice showed that various diets, even disturbances of short duration such as acute starvation or removing food for two days, resulted in acidic conditions. The important finding here is that it was not even necessary to induce starvation but to remove only an essential component from the diet to increase acidity and thereby influence susceptibility.[51]

Nutrition. Dubos entertained the widely accepted theory that poor nutrition increases susceptibility to tuberculosis. He described numerous nutritional regimens—from starvation to lobster and wine diets—that have been tried over the centuries to cure the disease. While acknowledging famine and pestilence often go together, his experiments tested whether changes in the nutritional state (not the nutritional regimen) toppled equilibrium from infection to disease.[52]

From 1940 to 1955, Rockefeller pathologist Howard Schneider headed a separate laboratory of experimental epidemiology that had shown diet could influence salmonella infection. He, like Dubos, found that the presence of a pathogen was not sufficient to cause disease. However, he attributed a diet's influence on disease to tiny amounts of chemicals in the food that became embedded in what he called the "fine structure" of tissues. The editors of *The Journal of Experimental Medicine* were troubled by Schneider's work. Dubos thought his population "statistics show just about nothing," and Rous complained that Schneider "was carried away by his ideas to the extent of being uncritical."[53]

Unlike Schneider, Dubos treated experimental diets in individual animals as nonspecific stresses. Without analyzing effects of individual nutrients or chemicals, Dubos and Cynthia Pierce demonstrated that the composition of diets markedly affected the survival of infected mice. For example, a diet of whole wheat and dried milk proved superior to one of cornmeal and butter, but the scientists cautioned that cornmeal did not necessarily cause a nutritional deficiency; just as likely, they concluded, it might have supplied a factor that enhanced disease.[54] Sudden food deprivation for short periods prior to infection also increased susceptibility to disease, whereas long-term deprivation or inadequate diets did not.[55] Even though mice were initially susceptible while on restricted diets, they recovered their resistance after they adapted to their deprivations.[56] When weight-reducing drugs were added to the diet, the animals became easily susceptible to several bacterial infections.[57] When diets differed in protein and amino acid concentration, mice survived longer on those containing more than twice as much casein. Interestingly, mice on both experimental diets gained weight at the same rate, thus confirming the troublesome fact, also found by other scientists, that diets adequate for growth and reproduction are not necessarily the best diets for resisting infection. A nutritional stress that increased susceptibility to disease could not be corrected simply by adding more protein; the protein needed a correct amino acid composition.[58]

An essential conclusion from this work was that indirect, nonspecific foodstuffs in the diet, or even one of its elements, could supply a toxin or could lack an essential nutrient; in other words, either kind of stress could kindle a silent infection into disease. While Dubos cautioned against blaming specific nutrients, he was adamant that changes in diet, mainly stressful and sudden ones, readily disturb one or more host mechanisms. Environmental factors related to diet are of course also important in disease, for wherever poverty, poor housing, crowding, and unsanitary conditions exist, then food is often scarce or of poor quality. The laboratory's more basic contribution was that nutritional stresses can bring about a collapse of an equilibrium in many other not so obvious ways.

Toxins. The laboratory also tested whether toxins that were well known for producing acute symptoms—from fever to tissue destruction, toxic shock, and death—could reawaken a latent infection. For nearly a century after Pasteur wondered about the role of bacterial toxins in causing disease, scientists have tried to determine their biochemical mechanisms. Dubos did not attempt to identify their nature or their lethal or protective actions, as he did with shigella toxins during the war work. Instead, whether toxins were derived from microbes, drugs, or environmental pollutants, he postulated that minute, innocuous doses of these biologically active substances would still manifest themselves, only more slowly and indirectly, to produce disease in hidden ways.

Several experiments showed the danger from exposure to low levels of bacterial toxins if an animal is simultaneously infected with a pathogenic microbe. When mice were injected with sublethal doses of toxins from either *Klebsiella pneumoniae*, pertussis vaccine, mycobacteria, or salmonella, there was no evidence of gross toxicity, at most only a transient weight loss. Following this exposure, the animals were challenged with virulent pathogens, such as staphylococci or mycobacteria. In this case, the outcome depended on the physiological state of the individual, causing death in a weakened animal, but sickness and eventual recovery in a healthy animal. In other experiments, the outcome depended on the time between toxin exposure and infection. Paradoxically, toxins could provide protection, also known as an antitoxic immunity, if given first. When toxin and pathogen were injected almost simultaneously, however, susceptibility was greatly increased and many animals died suddenly. When this experiment was reversed and toxins were given months after animals had been infected with sublethal doses of bacteria, there was an explosive multiplication of the dormant bacteria that resulted in death.[59]

Another toxin tested was cortisone, a commonly prescribed human drug; it produced even more alarming facets of latent infection. Rats and mice normally harbor a nonvirulent pseudotuberculosis, *Corynebacterium kutscheri*, and other scientists had shown that stresses from cortisone as well as deficient diets or irradiation could cause this quiescent and generally undetectable bacterium to reacquire virulence, multiply extensively, and cause death in previously healthy rodents.[60] When the Dubos group tested effects of cortisone in fifteen strains of mice, one strain that was highly resistant to most infections suddenly succumbed to "spontaneous" pseudotuberculosis from *C. kutscheri*, but the highly susceptible strain did not. As it turned out, the resistant strain had been harboring the infectious organism all along while the susceptible strain had not been a carrier.[61] Dubos often used this experiment to illustrate that animals (and humans) that are characterized as genetically resistant to a given pathogen may in fact be carriers of it. He alerted researchers and physicians to consider whether the drugs they test in animals or prescribe to patients may be the very source that provokes previously undetected and seemingly unrelated dormant infections to become active.

These original studies of latent infection established how very low levels of chemicals occupy a real ecological niche in pathogenesis, one not previously recognized. This research, along with Dubos' renewed warnings of antibiotic resistance in 1956, preceded by six years the publication of Rachel Carson's *Silent Spring* and Lewis Herber's *Our Synthetic Environment.* Their books alerted the public to the dangers inherent in thousands of recently introduced chemicals and initiated an outcry against environmental poisons and, in Carson's case, the banning of the pesticide DDT in the United States. Dubos, who regarded antibiotics as a kind of pesticide, always worried that continuous exposure to low doses of chemicals was far more ominous, calling them "pestilences that stealeth in the darkness." In contrast to the uproar over DDT and drugs that produced outright birth defects, allergies, and addictions, the warnings against antibiotic misuse did not arouse fears in physicians, patients, or the general public. Dubos criticized Carson and Herber for their "Panglossian attitude" in believing that ways could be found to retain the chemicals' advantage while effectively controlling their use and avoiding danger. This attitude, he said, "is like whistling in the dark. . . . All technological innovations, whether concerned with industrial, agricultural, or medical practices, are bound to upset the balance of nature. In fact, to master nature is synonymous with disturbing the natural order."[62]

Like many prophets before and since, he issued warnings of things we ignore at our peril: antibiotic resistance, hidden exposure to toxins, and emerging diseases. In retrospect, as he elaborated, human beings were simply accelerating these dangers by introducing novel pressures from chemicals in foods and drugs into natural processes and were moving toward what he originally feared in 1940: microbial worlds altered by these chemicals would then determine the pathology of many diseases.[63] One of his favorite sayings about relationships between the environment and human activities was Winston Churchill's argument for rebuilding the bombed English Parliament according to its original design. In a Dubosian paraphrase, we shape the diseases that afflict us, and afterwards they shape us.

These predictions are based on the simple premise—the crux of Dubos' experimental science—that "the microbe needs a fertile soil." The microenvironment, as revealed by his research, lies outside the germ theory of disease and traditional ideas of immunity. The germ theory, in his view, had become a "cult—generated by a few miracles, undisturbed by inconsistencies and not too exacting about evidence," whereas bacteria are only "opportunistic invaders of tissues already weakened by crumbling defenses."[64] Even in the presence of antibodies and cellular immunity, he further cautioned physicians to recognize that cells in a poor physiological state can no longer destroy microbes. "Immunity is necessary to prevent large fulminating infections but it is never sufficient to prevent reactivation of infection during the course of physiologic stress."[65]

Turning from Disease to Health

During this same period, René and Jean designed a country house on their Hudson Highlands property that reflected their vision of unobtrusively fitting into the rocky landscape. The first section was a four-room cottage, nestled against the boulders of Denning Hill in a grassy meadow above Old Albany Post Road, with walls of grey fieldstone salvaged from abandoned fences, expansive multi-paned casement windows, beams from the derelict two-hundred-year-old barn, low wooden gables, and a grey asphalt shingled roof. The interior walls were paneled with white walnut, which was discovered in a Vermont warehouse and fulfilled René's wish voiced twenty years earlier in his Greenwich Village apartment to be surrounded by wood. The large living room was dominated by a massive fieldstone fireplace and a grand piano that had once belonged to Marie Louise. There were few objects or books save for all of Thoreau's *Journals*. Soon afterwards, in anticipation of Jean's mother coming to live with them, they added a bedroom wing to the north, built of the same materials as the original, and playfully called "New North," the same name as his Rockefeller laboratory building. The basement of the new wing provided a one-car garage.

The two sections of the house were connected by a screen-enclosed breezeway. In Dubos' fantasy, this space was an orangery where he worked with Jean. On warm days it was completely open to the outdoor air. Jean, who did much of the literature research, read and wrote notes from books on her lap or edited a freshly written draft. René wrote tirelessly, encircled by sheets of yellow paper on the floor containing notes and drafts. She often referred to the books that took shape in this environment as "our children." Their pose recalls the famous enclosed-garden photograph of Pasteur working alongside his wife as they composed a scientific paper. Commenting on this photograph, Dubos thought it was good for Pasteur to work "with an orangery for laboratory, and trees and water for office furniture."[66]

Taking short breaks during the day, the Duboses walked through the woods, visited their "upstairs and downstairs" vegetable gardens, or watched frogs and snakes near the rock-lined pool. Dubos also tried some curious experiments that were his idea of fun. One year, he tried to devise a way to secure the scent of a wild spring fern. Another spring, seeds were coated with gramicidin to see whether his antibiotic might give plants a more robust start. Moss and rock gardens were copied from those he admired in Ireland and Japan. For a visiting cat, he played songbird recordings and marveled at its profound excitement. He was not discouraged when rabbits devoured the tiny cabbages and turtles feasted on the strawberries in the garden. A remark to his mother captures this lifelong *joie de vivre* and insatiable curiosity. "For me, the pleasure of gardening is in the effort to turn over the soil and not in the harvest." Perhaps this is why he brought pots of soil back to his city window ledge just to watch what might spontaneously emerge from the Garrison earth.[67]

In 1950, when the house was completed, the Duboses were featured in the premier issue of *Flair*, a short-lived, buoyant lifestyle magazine. For the photographer, René donned his tweed "Harvard" jacket and sported a walking stick to show off the open fields and rocks on his farm. For another photographer twenty years later, he donned the same odd jacket but carried pruning shears that were needed to tackle a returning forest and to rescue favorite trees. As *Flair* proclaimed, his "personal life represented a spiritual retreat into a garden of reason."[68]

The White Plague, the first of many books written in the orangery, is an unusual approach to medical history—a biography of a disease. It places tuberculosis in its social, economic, and cultural context and illustrates evolving views of the disease with fascinating stories from history, literature, and art. The book was prompted largely by his growing concern that understanding tuberculosis was not a bacteriological problem but instead "a social pathology."[69] The Duboses are critical of twentieth-century medical science for being more successful in retarding death with vaccines, surgery, and drugs than for finding other ways to protect against infection. They go as far as stating that no drug, however powerful and nontoxic, will eradicate tuberculosis and predict resistance to the newest (1952) antibiotic against tuberculosis, isonicotinic acid hydrazide (INH)—forecasts that remain valid today. Also eye-opening is their conclusion, "Tuberculosis will be conquered only when man has learned to function according to a physiological way of living that renders him more resistant to tubercle bacilli and when he has created a social environment that protects him from exposure to infection." This would not be difficult, they explained, because the ways of life affording the greatest protection against tuberculosis are "identical with the rules of good physical living . . . [and] well suited to fundamental happiness and health." *The White Plague* has remained in print for fifty years and endures as a medical classic.[70]

In 1952, just as the book was published, Dubos suddenly developed a massive ulcer that nearly killed him. Sensitized as he was to social factors in disease, he concluded about this illness that "worries as well as petty annoyances had a far greater influence on the vagaries of my duodenal ulcer than did dietary indiscretions."[71]

There were some major social and professional pressures that put him in this precarious medical condition. The beginning of several bold experiments, just described, coincided with the sudden transition of Rockefeller from an institute to a graduate university. From its early years, the scientific staff had been informally training postdoctoral scientists in all its laboratories, but now he worried that a formal doctoral program would counter the nature of this tightly knit cloister of twenty laboratories and dilute its intensity. In response to a letter from David Rockefeller, president of the board of trustees, Dubos suggested the Institute should lead the way with a novel kind of training in science. Since funding had increased substantially after the war, he anticipated there would be a "shortage of scientists with broad enough vision and training." Therefore, he proposed that exceptional individuals be selected and prepared to deal with "large theoretical problems which are often neglected merely because their solution is not in sight." Rather than being expected to "serve as helpers" on established research teams, they should be given the opportunity "to grow as independent thinkers and scholars." Echoing words of the retiring director Herbert Gasser, Dubos advised keeping the Institute "a place where scientific capital is being created, even though popular acclaim is more likely to reward those who spend the dividends of science."[72]

Instead, the Institute established a program for tutoring research fellows, leading to complaints from many members who believed the new president, Detlev W. Bronk (a biophysicist who had previously been president of the National Academy of Sciences and Johns Hopkins University), would have no sympathy for medical problems or broad biological views toward medical research. Dubos likewise feared the Hospital's influence in medicine would be weakened and that science would become abstruse and more reductionist.[73] His personal rebellion against this transition continued for several years and was articulated discreetly in his writing about the role of science in society, particularly *Dreams of Reason* and *Reason Awake*. His defiant stance was more apparent in his subsequent unorthodox experiments and in the unusually progressive research projects by the three students who trained in his laboratory on "large theoretical problems."

Other burdens were equally stressful during 1952. Dubos served as president of two major scientific societies, the Harvey Society and the Society of American Bacteriologists (now the American Society for Microbiology). Their administrative tasks were those he had previously avoided. The few reviews of *The White Plague* in science journals and news magazines were generally bland. Dubos became agitated that the book was a "flop" and deferred to Jean's opinion that their unusual thesis had not been well enough documented to rush it into print.[74] Waksman's review in *The New York Times* praised the Duboses' social treatment of tuberculosis but was critical that "fewer than two pages in the entire text are devoted to drugs, comprising the use of antibiotics, including streptomycin."[75] Dubos was no doubt further aggravated because Waksman had just won the Nobel Prize for the discovery of streptomycin as the "first antibiotic effective against tuberculosis," and especially so when *The*

White Plague stated that the disease "is not likely to be solved by the use of any drug."[76]

More personal disappointments followed. Jean's frail health was further compromised by the discovery of lesions in her lungs, posing the possibility that her tuberculosis was active again; this curtailed their travels and social activities. With his mother's death in 1954, he mourned his last connections to his French past. A year later, the death of his scientific mentor, Oswald Avery, further distressed him. Throughout this period, his ulcer remained a constant source of discomfort.

An obvious change came over Dubos. Laboratory colleagues noticed the long decade of moody episodes, during which they felt rebuffed by "the silent treatment" or his hypercritical comments. These moods were temporarily reversed when some new finding in the laboratory intrigued him or picked him up. Mostly, he kept to his office with the door closed. To his mother, he once had entertained thoughts of "whether it would be pleasant to be an animal that hibernates. To wake up each spring, new for the world, and the world new for it." Trying to think positively, he continued, there were some things that could transform themselves during hibernation and therefore it was better to not fight them.[77]

In 1955, during the depth of these agitations, historian Saul Benison conducted extensive interviews with Dubos for the newly organized Oral History program at Columbia University. Almost certainly, his reclusive behavior at the time colored his harsh comments about the changing Rockefeller environment and perhaps clouded or even altered the reminiscences he chose to relate about his early life. A remarkable aspect of this oral history is how Dubos spoke in paragraphs, entirely from memory, and for hours at a time with little prompting from Benison. To an outside reader, these pages provide a wonderful example of his storytelling ability, both charming and graceful. To readers of these pages who knew Dubos personally, including his wife Jean, these reminiscences of highly selective events and people tell only part of his life story until 1955. Considering his near-death experience due to the duodenal ulcer, it is worthwhile speculating that this oral history may reflect how, at that time, he wanted to be remembered. Indeed, many opinions expressed in 1955 changed dramatically during the following twenty-seven years.

Ecology of Human Health

Dubos' renewed search for health in the following years was greatly influenced by a chance reading of *The Way* (sixth century B.C.) by Lao-tzu, who associated violent emotional and physical changes of life with bodily and mental illness.

His search was undertaken with the "blind and naive ardor of the neophyte," by gathering examples from "wise men of all time" before the "germ" appeared in medical science, and by looking for illustrations of how healing and health were achieved empirically through wisdom, experience, and knowledge of human nature. His first lectures on the subject were tentative and cautious, testing the reactions of scientists and physicians who knew him as a rigorous bench scientist. He eventually concluded that giving "straightforward medical advice is not likely to bring about changes in the design of industrial equipment, the planning of cities, or in smoking habits," but perhaps "a semi-religious-philosophical doctrine could do it."[78]

Some lectures had romantic titles such as "Microbiology in Fable and Art," "The Philosopher's Search for Health," and "The Gold-Headed Cane in the Laboratory." Others were provocative and argumentive, including "Second Thoughts on the Germ Theory." Still others, like "The Evolution and Ecology of Infectious Diseases," were more reasoned. All were preliminary essays that became the basis of his most influential and popular book, *Mirage of Health*.[79]

He arrived at a definition of health by posing a series of paradoxical questions: *Is health achieved or restored by treating disease? Is health acquired or preserved by following ways of nature? Is health in the realm of experimental science or clinical art?* Weighing evidence from laboratory experiments showing health is an equilibrium in which everyone harbors disease germs, but not everyone is sick, he adopted "the Hippocratic view that health is universal harmony, and that the role of the physician is to restore equilibrium between the various components of the body and the whole of nature."[80]

With these questions, Dubos anticipated the perils of a medical philosophy that ignores the evolutionary consequences of its behavior. He continued to reject the common militant scientific view that microbes are aggressors, the body mobilizes its defenses, and the physician fights to eradicate infection and conquer disease. Instead, the web that connects all things on earth holds together through making peace rather than war, and through health rather than disease. On one side, he noted evolutionary drawbacks to modern medicine that were so thoughtlessly developed, so powerful, and so indiscriminate that they would *accidentally* influence the "future of the human race," even "alter significantly the constitution of the human stock." It was possible these new techniques of prophylaxis and therapy would interfere haphazardly "with the processes of selection that upheld in the past the genetic quality of the human race." Expanding on the idea of using antibiotics to eradicate any disease, he advised this would leave open the possibility that natural selection and back mutations would return whole populations to a naive state of devastating susceptibilities to pathogens, foreseeing perhaps those organisms synthetically created and willfully spread by weapons of black biology.[81]

On another side, Dubos saw some evolutionary advantages of latent infections. He objected to Darwin's view of a "struggle for existence" as unconcerned with microbial life and so imbued with competition that it did not consider cooperation. To most people, Darwinian success implies that an organism can destroy or master its enemies. In contrast, Dubos held that peaceful coexistence of pathogens and hosts is an ultimate stage of prolonged contact that is beneficial and even vital to both organisms. Accepting this scope of evolutionary thinking also allowed that periodic struggles for existence (disease) would occur and might even become biological necessities at times of extreme stress.[82] Unsurprisingly, English biologist Julian Huxley—a promoter of the Modern Synthesis that the gene is the unit of natural selection—read *Mirage of Health* in manuscript and objected to Dubos' views, to which Dubos responded that "the theory of evolution as formulated by zoologists fails to incorporate many of the facts which become limiting factors in medicine."[83] More forcefully, he observed elsewhere, "I would be surprised if the medical geneticist of the future did not rate the sociological aspects of his science as more important than its contribution to biochemistry."[84]

There was no room in Dubos' arguments for what he called medical utopias. "Medical science permits, and ethics demand, that all children survive, however defective they may be physically or mentally, . . . and reproduce because they are provided with continued and expensive medical care." Genetic defects and physiological deficiencies, he added, need not be a serious handicap in a society equipped to deal with them. As discussed in chapter 6, Dubos championed the view that humans can live and function effectively even though they are tuberculous, blind, diabetic, crippled, or psychopathic. However, he warned, a time may come when an accumulation of the weak and sick, who depend on such technological medical care in these seeming medical utopias, will create such a heavy social burden that some aspects of medical ethics would have to be reconsidered in the harsh light of economics. Prophetically, he fathomed, "There is no way to predict the distant outcome of this state of affairs because it has no precedent in the life of man."[85]

With his book *Mirage of Health* (1959), Dubos toppled two illusions: the quest for perfect health and the conquest of disease.

Health, in his definition, "is not necessarily a state of well-being, not even . . . a long life. It is, instead, the condition best suited to reach goals that each individual formulates for himself."[86] There is an unusual emphasis of negative terms in his expanded definition: health is not an absolute ideal, not a utopian goal, not purchasable, not measurable, and not achieved by pursuing one palliative and protective measure after another. Health cannot be defined in terms of anatomical, social, or mental attributes. Two reasons might be suggested for this rhetorical device. It helped divorce his meaning of health from medicine and it reinforced his argument that the responsibility for achieving health belongs with the patient, not the physician or scientific medicine.

Dubos eschewed the idea of complete well-being and observed that imperfect, even diseased, humans are capable of achieving a rewarding existence in an imperfect world. This meaning contrasts with the definition of health in the World Health Organization's charter as "a state of complete physical, mental, and social well-being and not merely the absence of disease or infirmity." He argued instead that "the earth is not a resting place," and human life is a dynamic process, so to renounce adventure, abandon risk, or avoid the world would result in a static paradise. Once we accept that complete freedom from disease "is almost incompatible with the process of living," then we will know our aspirations cannot be satisfied with an easy life or perfect health.[87]

Two memorable analogies illustrated why the quest to conquer or eradicate disease is also futile. From a biological perspective, Dubos compared infectious diseases to weeds: when one species recedes, another advances to take its place. From a social perspective, he compared the use of antibiotics to the "naive cowboy philosophy that permeates a wild west thriller." The death of the villains (pathogens) does not solve the fundamental problem, "for the rotten social conditions which opened the town to the desperadoes will soon allow others to come in, unless something is done to correct the primary source of trouble." He blamed the magical power of antibiotics for blunting the critical senses of physicians, scientists, and laymen alike, because "men still find it easier to believe in mysterious forces than to trust in rational processes." Instead, antibiotics should be considered short-term solutions to acute situations and not the long-term answer to the perennial problem of infectious diseases.[88]

Mirage of Health pioneered a basic shift in medical thinking about health. At the time, these ideas seemed revolutionary, although today they are nearly taken for granted, even widely accepted by medical educators, physicians, and scientists, many of whom came to regard Dubos as "the conscience of medicine."[89] Recently, Daniel Callahan, a leader in biomedical ethics, echoed these sentiments in his book *False Hopes*, calling Dubos "absolutely right." "The mirage shimmers," Callahan noted, "because it basks in the sun of scientific optimism and often-desperate hope." However, it is a "false hope" for medicine to master illness and conquer infectious diseases.

Parents of René Dubos, Georges Alexandre and Madeleine Adéline De Bloedt Dubos, circa 1908. Courtesy of Francis Dubos.

Hénonville, France, Route de Pontoise, 1909.
Courtesy of Francis Dubos.

In 1907, the French village of Hénonville had 450 inhabitants, including six-year-old Dubos, fourth from right in front of the Café de la Gare.
Courtesy of The Rockefeller Archive Center.

René Dubos at left, circa 1908, at age seven, Hénonville, France. His mother is standing, brother Francis is on the knee of an unnamed seated woman, and sister Madeleine is standing on right. Courtesy of The Rockefeller Archive Center.

Clockwise from left:
René Dubos, Collège Chaptal, 1913. Courtesy of Francis Dubos.

René Dubos, Rutgers University, 1924. Courtesy of The Rockefeller Archive Center.

Vial holding type III pneumococcal polysaccharide labeled in Avery's handwriting. Courtesy of Medical World News Collection in the Houston Academy of Medicine-Texas Medical Center.

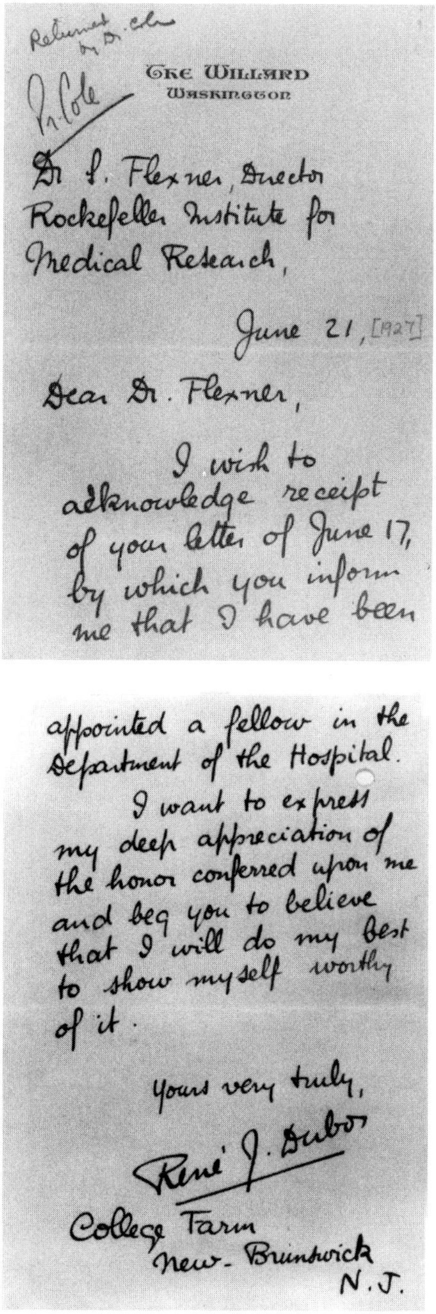

Returned by Dr. Cole

Dr Cole

THE WILLARD
WASHINGTON

Dr S. Flexner, Director
Rockefeller Institute for
Medical Research,

June 21, [1927]

Dear Dr. Flexner,

I wish to acknowledge receipt of your letter of June 17, by which you inform me that I have been appointed a fellow in the Department of the Hospital.

I want to express my deep appreciation of the honor conferred upon me and beg you to believe that I will do my best to show myself worthy of it.

Yours very truly,

René J. Dubos

College Farm
New-Brunswick
N.J.

Two-page letter dated 21 June 1927 from René Dubos accepting Simon Flexner's offer to join the The Rockefeller Institute (reduced). Courtesy of The Rockefeller Archive Center.

René Dubos with Jacob Lipman, Selman Waksman, visitors, and fellow students of the New Jersey Agricultural Experiment Station at Rutgers University, 1927. Seated, left to right: Mr. King, René Dubos, Edward H. Folwell, Stavano B. Fagundes, Mr. Muller, Dr. Blom. Standing, left to right: D. J. Hissink (Secretary General of International Society of Soil Science, with watch chain), A.W. Blair, Selman Waksman, Dr. McLean, Jacob Lipman, Robert Diehm, Edward J. Russell, Elias Melin, Henry Adams, Mr. Reuszer, Robert L. Starkey, Arthur L. Prince, I.V. Shunk, Florence Tenney. Photograph by George H. Pound. Courtesy of American Society for Microbiology Archives.

Oswald Avery's respiratory disease laboratory group during the early nineteen-thirties. Seated, left to right: Thomas Francis, Jr., Oswald Avery, René Dubos. Standing, left to right: Edward E. Terrell, Kenneth Goodner, Frank H. Babers. Photograph by Walther F. Goebel. Courtesy of The Rockefeller Archive Center.

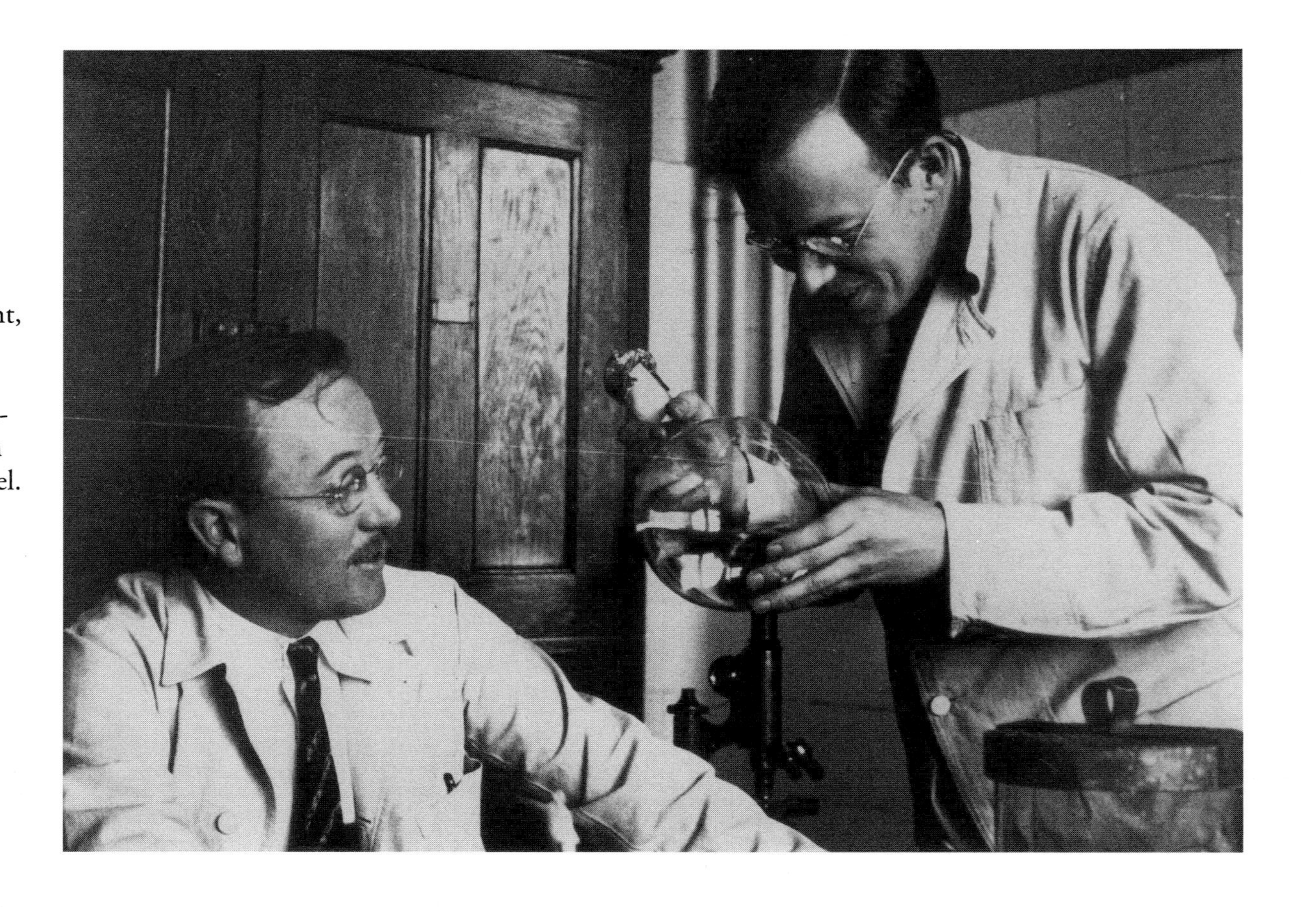

René Dubos on right, showing flask to Kenneth Goodner, during the nineteen-thirties. Photograph by Walther F. Goebel. Courtesy of The Rockefeller Archive Center.

[Reprinted from THE JOURNAL OF EXPERIMENTAL MEDICINE, July 1, 1939,
Vol. 70, No. 1, pp. 1–10]

STUDIES ON A BACTERICIDAL AGENT EXTRACTED FROM A SOIL BACILLUS

I. PREPARATION OF THE AGENT. ITS ACTIVITY IN VITRO

BY RENÉ J. DUBOS, PH.D.

(*From the Hospital of The Rockefeller Institute for Medical Research*)

(Received for publication, April 17, 1939)

Microorganisms perform a vast number of biochemical reactions, many of which are not known to occur in the animal and plant kingdoms (1). On the basis of present knowledge it is conceivable that one may find in nature microbial species endowed with catalysts capable of activating almost any type of biochemical reaction. During the past few years, this point of view has found its application in the isolation of soil microorganisms which selectively attack certain substances of interest to the biochemist (2) and to the immunologist (3–8). It may be recalled in particular that soluble polysaccharides, extracted from several bacterial pathogens, have been found to be decomposed by certain microbial species, although the same substances are resistant to the action of all known enzymes of animal and plant origin.

It appeared possible that there also exist in nature microorganisms capable of attacking not only isolated soluble components of other bacterial cells, but also the intact living cells themselves. Actually we have isolated from soil a spore-bearing bacillus which attacks and lyzes the living cells of several species of Gram-positive microorganisms. The present paper describes the isolation of this new soil bacillus, and the preparation, properties, and activity of the soluble agent by means of which it attacks and lyzes the living cells of the susceptible, Gram-positive species.

EXPERIMENTAL

Isolation of a Sporulating Bacillus Capable of Lyzing the Living Cells of Gram-Positive Microorganisms.—The method employed for the discovery of microorganisms capable of attacking certain definite organic compounds has already been described (2, 3). It is based on the assumption that all organic matter added to the soil eventually undergoes decomposition through the agency of microorganisms. In the present case, it was hoped

The detailed report describing the discovery of the antibiotic tyrothricin. Reproduced from *The Journal of Experimental Medicine*, 1939, volume 70, page 1, by copyright permission of The Rockefeller University Press.

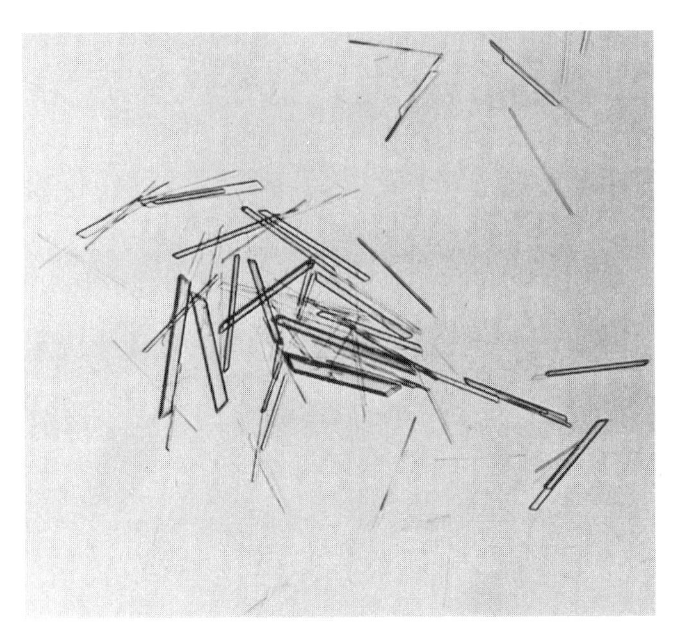

Tyrocidine hydrochloride crystals, 1939.
Courtesy of Dr. Rollin D. Hotchkiss.

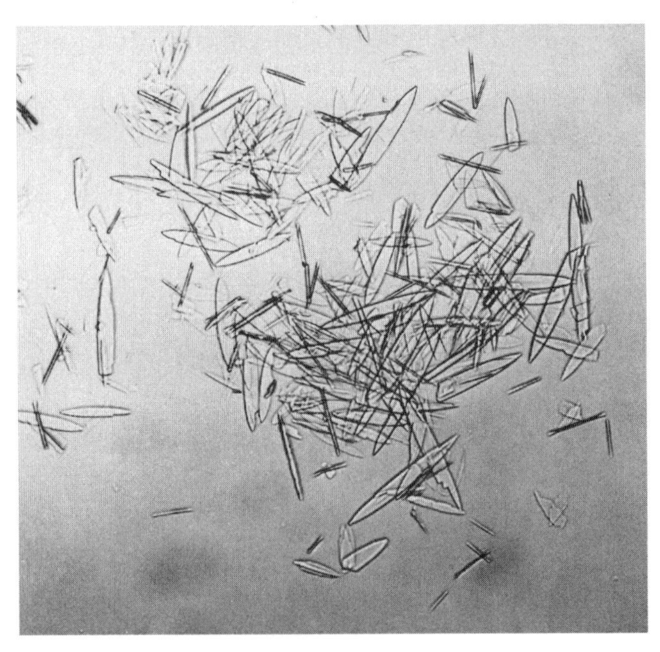

Gramicidin crystals, 1939.
Courtesy of Dr. Rollin D. Hotchkiss.

Rollin D. Hotchkiss, 1935.
Courtesy of The Rockefeller Archive Center.

Marie Louise Bonnet Dubos, after 1934.
Courtesy of The Rockefeller Archive Center.

Marie Louise and René Dubos, Ray Brook sanatorium, circa 1941.
Courtesy of The Rockefeller Archive Center.

Letha Jean Porter, circa 1941. Author's collection.

Dubos laboratory group at The Rockefeller Institute for Medical Research in 1946. Back, left to right: Jane Kramer, Bernard D. Davis, Alfred Marshak, René Dubos, Merrill Chase, Cynthia Pierce, Rollin Hotchkiss. Center, left to right: Margaret Brophy, Harlean Cort, Gardner Middlebrook, Anne Christiansen. Front, left to right: Jane Buckalew, unknown, Hannah Schottlander, Mary Andrews. Author's collection.

Jean and René Dubos, 1950.
Photograph courtesy of Roger Greif.

James Hirsch, Jean Dubos, René Dubos, and Zanvil Cohn in 1971
at the Dubos seventieth birthday festschrift celebration.
Photograph ©Lawrence R. Moberg.

René Dubos giving the Jacques Parisot Foundation Lecture, "Human Ecology," under the auspices of the World Health Organization in 1969. Courtesy of The Rockefeller Archive Center.

The Dubos house in Garrison, New York, 1972. The breezeway between the two sections was favored for writing from early spring until late autumn. Photograph ©Lawrence R. Moberg.

René Dubos in his Rockefeller University office, 1972. Courtesy of the National Archives II-Still Picture Division, RG 16, Series NF, College Park, Maryland.

René Dubos at his house in Garrison, New York, 7 April 1972.
Photograph ©Lawrence R. Moberg.

Grave site of René and Jean Dubos,
St. Philips in the Highlands, Garrison, New York.
Photograph ©Lawrence R. Moberg.

Callahan is not alone in putting the very quest for these two objectives at the core of today's "health" care crises. These and other long-neglected concerns of evolutionary medicine are now becoming painful lessons and major topics of interest in infectious disease.[90]

Pursue Well-fare not Warfare

With *Mirage of Health*, Dubos became the leading medical scientist to oppose the current popular belief that infectious diseases could be conquered. He argued that science could theoretically solve biological aspects of almost any medical problem created by the rapidly changing social and technological order, but limited economic and human resources would make this goal practically impossible. He criticized the public for becoming too confident in science and urged audiences to select which medical problems had priority and to determine the kind of health they want. He warned these issues would cause ethical dilemmas requiring difficult choices "made by society as a whole because they will involve value judgements. More and more medical science will need to be integrated into social conscience." As a result, for a true era of social conscience to arise, society needs to make sense of health rather than medicine.[91]

More specifically, in a classic lesson delivered to countless medical students, Dubos anticipated a different future for infectious diseases. It involved seeking a more peaceful coexistence with pathogens, embracing their complexity, respecting their evolutionary drive, and launching ways to increase our tolerance of them. The lecture also outlined scientific strategies, whose research base has yet to arrive, to anticipate epidemics that were certain to occur. These strategies included ways to supersede herd immunity and infection immunity, in other words, to replace accidental exposures to pathogens with a controlled colonization of microbes that would furnish lifelong immunity. Using proper techniques, supplemented by judicious use of antibiotics, vaccines, and sanitary measures, "it should be possible to establish in all normal human beings a high degree of tolerance to the many pathogens that will inevitably continue to be part of man's environment."[92]

Dubos commonly opened this lecture with a provocative question, "Do you need to take antibiotics every time you get sick?" He sometimes cited his own two bouts of subacute bacterial endocarditis and how massive intravenous doses of antibiotics saved his life. After ending the lecture, he answered this question just as provocatively. "First, think about where and how your disease arose. Take antibiotics only if it's absolutely necessary. Then," he prescribed, "change your life."

By 1960, Dubos had moved from microbiologist to macrobiologist and replaced notions of antibiosis with symbiosis. In his evolving views of scientific medicine, he realized that the task of achieving health would demand a kind of wisdom and vision transcending scientific knowledge of disease. During the 1960s, the years described in the next chapter, experimental pathology was relinquished in favor of environmental biomedicine as he began to advocate a new science to generate evidence of environmental influences on health and disease. A new hypothesis also emerged: to improve health, it is first necessary to understand our impact and interactions with our surroundings. The time had come to show that humans do not live in a biological vacuum.

5 Toward a Science of Human Nature

Our life is not all moral. Surely, its actual phenomena deserve to be studied impartially. The science of Human Nature has never been attempted, as the science of Nature has. The dry light has never shone on it. Neither physics nor metaphysics have touched it.

HENRY DAVID THOREAU

IN 1962, AT THE AGE OF 61, Dubos published a pithy book that was read by few and reviewed by no one. This small volume called *The Torch of Life* was written, in his words, "from the very marrow of my bones" during an obsessive six weeks the preceding hot, steamy summer in Garrison. Its flowing thoughts are acutely self-aware and full of experiential passages, yet the pages lack personal details.[1]

The Torch of Life is an intellectual autobiography, the credo of a biologist and humanist. It can be read today as a veiled response to his professional and personal struggles during the 1950s. More revealing is how its incompletely sketched ideas frame his future searches. Definitions of health that were phrased in negative terms in *Mirage of Health* are replaced here by constructive views of health as a potentiality and a creative adaptation.

A key to understanding Dubos' maturing philosophy appears in these highly telescoped and interconnected statements. Two anecdotes in *The Torch of Life* suggest where his ambition might lead and both relate to feats of construction. The opening pages evoke Washington Roebling's physical collapse under the crushing load of finishing his father's Brooklyn Bridge. Though paralyzed, he supervised the construction of this "poetical masterpiece" from his bedroom window. Dubos remarked how this task showed "man at his highest—more concerned with worthy creation than with health and comfort." The closing pages relate the fable of three medieval hod carriers on the road to Chartres: one says he is carting stones, another says he is working on a wall, and a third says he is building a cathedral. Dubos concluded that the mysterious sense of responsibility toward the future expressed by the third carrier is what makes so many men willing to work for causes that transcend their selfish interests. No matter how "limited their individual strength, small their contribution, and short their life span, their efforts are never in vain because, like runners in a race, they hand on the torch of life."

Set down between these anecdotes are many ideas, or building stones, that Dubos would shoulder the rest of his life. He could not have known whether his own task ahead would be carting stones, working on a wall, or building a monument to leave behind. "As long as you think 'in bricks'," said Nobel laureate in chemistry Ilya Prigogine, "the bricks can last for thousands of years. But if you think 'in cathedral,' there is the moment when it is constructed and the one when it falls in ruins. And with the same bricks, you can construct other cathedrals and palaces."[2] Dubos formed many "bricks" in this book from the knowledge he had accumulated during thirty years in scientific medicine. During the next two decades, he used and reused these "bricks" to structure many concepts that he believed would lead to a better understanding of human nature.

Dubos always regretted that the book was part of "The Credo Series" in the company of spiritual autobiographies by leading scientists, philosophers, theologians, and political thinkers. *The Torch of Life* was contemplative but free of religious creed. Although he was raised a Roman Catholic and was affected by the church during his youth, he observed late in life, "even though I have been a bad Catholic and have disagreed with many policies of the church, I have retained an emotional allegiance to it."[3] He was slightly involved, during the 1950s, with the American Friends of Saint-Benoît-sur-Loire—where Benedict is buried—because of his interest in the Benedictine approach to land stewardship.[4] Dubos remained steadfastly indifferent to any formal religious ritual and doctrine, saying "I am a poor Christmas Christian but a devoted believer in rebirth at Easter," and he never joined a religious institution.[5] It was entirely expected of course that a biologist would ask "What is Life?" To Dubos, the answer to this question "always implies something immaterial. There is a great mystery and I believe there are such mysteries that go beyond human understanding. Although I feel there is something religious in me, I have never succeeded in formulating it." He nonetheless attached significance to the Latin word *religio*, with its root *lig-* (to bind or unite), for conveying "the poetic sense that to be human is to be linked to all the rest of creation." Religion to him was simply "a force of profound humility, for we cannot explain the most important things of the human spirit."[6]

Instead, the spirit of the book reveals the special French persona of Dubos as *essayiste*, *provocateur*, or *intuitif* who postulates ideas without prescribing or consciously creating a system or philosophy. He had previously decided it was "unrewarding for a philosopher to demonstrate his thesis with too much thoroughness and too convincingly. His ideas soon become part of the intellectual household of humanity and . . . either are forgotten or their memory becomes somewhat boring."[7] His self-characterization was based on a statement by eighteenth-century French philosopher Jean Jacques Rousseau that he would rather be known for his paradoxes than his prejudices. Dubos often used paradoxes, never displaying only one aspect of an idea, always exploring contrasts, leaving the topic for others to expand or conclude. So many of his ideas became so prominent that his friend, French Nobelist André Lwoff, writing in 1965, thought that if Nobel Prizes were given for "ideas rather than molecules" then Dubos should have received one years before.[8]

The unifying thought in *The Torch of Life* came from a riddle that challenged him throughout his life: "The final questions of science as of philosophy are not about matter and energy, but about the nature of life in a universe where lifelessness is the rule, life the puzzling exception."[9] This book was the starting point for organizing intuitive ideas about how to ferret out many of these puzzling exceptions. Ideas that appeared first in this book—beast or angel, quality of life, biological partnerships, logical versus willed future, health as adaptability, the pursuit of happiness, and the open future—eventually became themes of a dozen books and hundreds of lectures during the next two decades.

The experiments and essays during the 1960s are perhaps the least known and understood yet the most revolutionary and formative of Dubos' career. In the laboratory, there were novel, assertive experiments to test an organism's responses to environmental influences. On the lecture circuit, he tested ideas and asked provocative questions about human responses to environmental influences, a process he described as trying to create a science of human nature.[10] Underlying both activities were concerns that science had failed to recognize the uniqueness of human life: "We do not know what we are nor where we are, what we want to become nor where we want to go."[11] Dubos was careful not to make dogmatic claims about the evidence from his uncommon research models, and he was not bothered by inconsistent or incomplete ideas in defining a new science. Everything during this period propelled his evolution from a bench scientist concerned with experimental results to a humanistic biologist concerned with values.

Laboratory Research in Environmental Biomedicine

Consistent with an enlarging ecological perspective, Dubos' experimental systems had broadened substantially during his career—from soil microbes, to bacterial pathogens, to host tissues, and, during this final stage, to organ systems and organisms. This career did not progress by accumulating facts and following a straight, endless road, marked at intervals by milestones. The unifying nature of his scientific creativity, most obvious at this stage, was not discoveries of particular isolated facts but finding relationships; at each stage these discoveries opened distinctly different ways of looking at nature and caused him to change scientific directions. The experiments in this final phase focused on the relation of the total body to the total environment after he decided it was defeatist to pretend that some complexities of human beings could not be studied experimentally until their submicroscopic structures and chemical processes were defined. This approach was decidedly outside traditional science and indicates his freedom to act on his belief that "the very success of the reductionist approach has led to the neglect of some of the most important and probably the most characteristic aspects of human life."[12]

A symbol of this new phase was the laboratory's name change in 1961 from Bacteriology and Pathology to Environmental Biomedicine. The prefix *bio-* was added so as not to imply the laboratory was studying human patients or that Dubos was a physician. The name also placed the research outside the traditional medicine laboratories associated with the Rockefeller Hospital. "Environmental" and not "ecology" was chosen because he routinely called ecology "the vaguest word" in the English language and criticized it as a science that ignored the role of human beings. Moreover, pathologist Howard Schneider had named his laboratory "Ecology," defining it as medical ecology or epidemiology, which took a genetic and nutritional approach to disease by creating model epidemics in populations of mice.[13]

Despite these practical considerations, the main inspiration for the name came from Hippocrates' perception that the environment influences human health and the physician's art. Hippocrates had described the relevance of such environmental forces as climate, topography, soil, food, and water to phenomena of health, diseases, medicine, and even sociology and military prowess. Although these ideas were at the "birth of environmental medicine," Dubos said, they ceased being a general guide for physicians at the end of the nineteenth century when two advances "made Hippocratic medicine obsolete, namely the doctrine of specific etiology, and the description of the organisms in terms of elementary structures and mechanisms."[14]

Specifically, the laboratory studied two systems where environmental factors were critical: the normal working bacterial organisms, or biota (in Dubos' term, flora), in the gastrointestinal tract, and neonatal development. Even after forty years, it may be too soon to judge these pioneering studies. Dubos knew the work was neither understood nor appreciated by many scientists, including his closest associates and laboratory successors. James Hirsch claimed this research was "not worth a hill of beans," and Zanvil Cohn did not consider the research "hard science." At the time, they were puzzled why Dubos deliberately embarked on untried areas and strayed farther from research ingrained with reductionist traditions, although much later they recognized the larger implications of these studies.[15]

Of historical interest, these studies received partial support from the National Institutes of Health. It goes almost without saying that this hypothesis-driven, innovative science would not fare well in today's competition for funding. Until 1956, The Rockefeller Institute had not allowed outside funding and financed its research solely from its endowment. As summarized by Gasser, grant projects "by definition are not consonant with free inquiry" because they compromise the initiative and independence of the scientist.[16] However, since Dubos was well enough established and the federal grant system was still fluid, he was able to take risks and devise bold experiments that helped enlarge his thinking about health.

The digestive tract as an ecosystem. Before the mid-1950s, many physicians believed that the stomach and small intestine contained only those bacteria ingested with food. By extension, the presence of bacteria, particularly the aerobic *Escherichia coli* and enterococci, meant there was infection or disease. Dubos moved beyond work by contemporary scientists who limited their experiments on the digestive tract to the easily reached mouth, throat, and rectum. As a result, the laboratory discovered that not only does the gut normally harbor a large variety of microbes but their integration in bodily activities is important for maintaining health and preventing disease.

The working hypothesis was that the digestive tract "constitutes a highly integrated ecosystem" where each component behaves independently but is influenced by the others and by environmental factors.[17] Just as Winogradsky discovered a native microbiota in soil that had been overlooked by experimenters using artificial culture media incubated in air, Dubos assumed the gastrointestinal tract also harbors a vast indigenous population of microbes that had not yet been suitably isolated.

As a result of developing new culture techniques, the Dubos laboratory discovered many previously unseen and uncultivatable species of naturally occurring microbes along the gastrointestinal tract. With Dwayne Savage, Dubos went on to establish precise histological techniques that allowed them to identify specific bacteria in distinct locations: the bacteria each have exacting growth requirements, assume a regular pattern of colonization in a newborn animal, and associate with epithelial surfaces to establish a succession of microenvironments along the tract.[18] Other important findings revealed that these microbes are acquired soon after birth and persist in constant numbers throughout an animal's life, most of the bacteria are anaerobic, and areas previously assumed to contain mostly dead or pathogenic bacteria in fact normally harbor large populations of living organisms.[19]

These observations led to their dividing the microbial populations in the digestive tract into three categories following a terminology similar to Winogradsky's descriptions of soil ecosystems. *Autochthonous*, or symbiotic, organisms are those present during a long period of evolutionary association with the animal species and essential for development and function. The *normal*, or ubiquitous, microbiota become established in practically all members of a community but can vary in population size from one community to another. *Indigenous* microbiota include the autochthonous and normal organisms, plus the pathogens acquired accidentally throughout life that remain in the body. The indigenous population strongly supported Dubos' earlier thesis: diseases that seem to occur spontaneously are not necessarily genetic in nature but can result from something disturbing this equilibrium and allowing the resident yet quiescent pathogens to become active.[20]

Even with this evidence in hand, Dubos was quite sure that the newly identified organisms were still a small sample of what actually exist. Subsequently, research by others has found at least four hundred species of bacteria that regularly colonize our bodies, most still uncultivatable and unidentified below the genus level. A continuing challenge of a gastrointestinal microecology, as Dubos advised, depends on eliciting the full ecosystem of bacteria and, more relevantly, learning what factors upset its equilibrium and influence whether the host is sick or well.[21]

In tests of two environmental influences—antibiotics and diet—on this ecosystem, the laboratory went on to demonstrate that the gastrointestinal tract cannot develop fully or function normally unless it harbors a certain microbial biota.

As expected, antibiotics eliminated so many necessary intestinal bacteria when they were given to test animals that an uncontrolled growth of the nonsusceptible bacteria caused unusual diseases, the nature of which depended on past infectious experiences of the animal. After antibiotics were discontinued, the microbiota returned to their initial levels, but only if the animals were fed proper diets. In other experiments, feeding antibiotics to animals caused surprising weight gains by reducing or removing microbes that competed for nutrients but at the same time opened the way to secondary infections or vitamin deficiencies. Incomplete as these experiments were, they substantiated the well-documented clinical experience in which antibiotic therapy is often followed by infections in various organs by microorganisms usually considered of low pathogenicity.[22]

In experiments with diet, mice having a normal microbiota were compared with two special breeds: one strain was developed at Rockefeller to exclude certain pathogens and gram-negative bacilli and the other was a strain of germfree mice that lacked microorganisms. Tests on those with modified bacteria showed that the composition of microbes in the intestinal tract profoundly influences the animal's ability to utilize food as well as to resist various infections and lethal toxins. Tests with germfree mice showed that their lack of microbiota also altered their nutritional requirements, lymphoid tissue, circulatory physiology, and composition of serum proteins. Importantly, these conditions in the special breeds were not the result of genetic mutations, as some observers suggested, because they could be corrected or reversed, thus reverting to the biota of normal mice, either from being nursed by normal mothers or by changing their diets.[23]

An unexpected confirmation of many previous experiments on stress and disease occurred when two air conditioners failed in the breeding room of these special mice. The sudden temperature increase caused a profound alteration in the composition of their intestinal biota and increased their susceptibility to disease. Dubos often singled out this unplanned incident as another example of how certain traits assumed to be genetic were instead produced by the environment.[24]

Another series of experiments showed that normal microbes in the digestive tract are needed to resist pathogens. Rose Mushin and Dubos tried to establish foreign bacterial species, such as the pathogen *E. coli*, in the gastrointestinal tract of mice, but they were initially unsuccessful, as the native microbes rapidly eliminated this invader. However, by giving antibacterial drugs before the experimental infection, *E. coli* immediately colonized the tract in the absence of the normal biota. Strictly speaking, the indigenous microbes were not classified as a host resistance mechanism since the microbes are outsiders, albeit symbiotic or living with the host. Instead, their role is one of interference to produce substances or create microenvironments that prevent alien invaders from taking up residence.[25]

The evidence that resident microbiota are symbiotic with their hosts and function as an ecosystem, Dubos concluded, had demonstrated the necessity of living peacefully with microbes. The microbes can influence growth rate, development, and function of tissues, which foods are necessary, and how they are used. The microbes also control whether transient pathogens colonize or upset the ecosystem. Their composition is not fixed and can change profoundly, qualitatively and quantitatively, as they continuously interact in response to environmental and internal factors. These interactions in turn can determine which substances are released from the digestive tract into the general circulation. Moreover, when their equilibrium is greatly disturbed, forces are set in motion that can impinge on health and disease. The microbial biota, he said, are integral to the normal development and function of men and animals "to such an extent they cannot survive in the open world without them."[26]

Environmental influences, or biological Freudianism. William H. Stewart, Surgeon General of the United States, proclaimed in 1969 that the war against infectious diseases had been won. A decade earlier in *Mirage of Health*, Dubos had argued just the opposite, that disease is inevitable and its control "is a never-ending task that demands eternal vigilance." He would not have been surprised today that old infectious diseases are raging anew and that as many as thirty previously unknown scourges have emerged since 1970, among them Ebola, Lyme, AIDS, and mad cow disease. Many people now want to attribute the source of these diseases to drug-resistant forms of bacteria, the appearance of seemingly new pathogens, and the failure to apply with sufficient vigor the existing knowledge of prophylaxis and therapy. Before 1960, Dubos had been promoting an ecological strategy to keep microbial habitats in balance, to anticipate their evolutionary drive to domesticate us, and to focus on prevention not defense. During the 1960s, he enlarged this approach and advised looking elsewhere for the real source of diseases to come—our personal lifestyles.

An important hypothesis to emerge during this time was that unconsciously adapting to environmental factors can be dangerous. In *Man Adapting* (1965), there is considerable scientific evidence describing the effects of misguided over- or underadaptation to modern ways of life. Dubos adopted the phrase "diseases of civilization" to convey how human diseases throughout history had reflected changing social patterns and how previously unknown diseases in the future would result from inevitable shifts in diets, drug use, social circumstances, technologies, and ethics. In addition, he warned of new biological dangers to health, including exposure to noise, air-conditioning, radiation, air and water pollutants; estrangement from natural cycles of time, seasons, and climates; and emotional trauma from crowding, excessive stimuli, boredom, featureless surroundings, compulsory leisure, and automation. Most worrisome, the price of adapting biologically to these seemingly benign stimuli could not be predicted, he explained, since they are unknown in our evolutionary past.

A corollary to topics in *Man Adapting* was a final set of laboratory experiments on "lasting effects of early environmental influences." *Early influences* has been commonly used by the social sciences to indicate conditioning of behavior. Dubos introduced the startling concept *biological Freudianism* as a physiological counterpart to Freud's emphasis on damaging emotional experiences in early life. Many scientists had determined that newborn animals are highly susceptible during formative stages to such influences as nutrition, infection, temperature, crowding, isolation, or a variety and intensity of stimuli, and that any insult experienced by the mother, such as the extreme human tragedy of thalidomide, affects her offspring. The Dubos laboratory added to this research by originating a model to test stimuli at such innocuous levels that their initial effects were imperceptible and to test them during the poorly understood prenatal and early postnatal periods of life. A working hypothesis was that mild insults are indirect, cumulative, and can produce harmful effects that are long delayed and irreversible.[27]

During a 1967 keynote speech on human development at the National Institutes of Health, Dubos argued that for almost any human problem it is possible to develop experimental situations in animals. He presented the laboratory's working model "to show that many problems of the underprivileged populations can be reproduced in the laboratory by exposing newborn mice to nutritional deficiencies, subclinical infectious processes or behavioral disturbances not unlike those commonly experienced by deprived people."[28]

The laboratory model addressed a situation witnessed by Dubos in an underprivileged human population. On several visits with physician-scientist Leonardo Mata to highland villages in Guatemala during the 1960s, he observed the devastating role of infection leading to high rates of infant morbidity and mortality in these Indian populations. Scientists and physicians from the Institute of Nutrition in Central America and Panama and WHO's Pan American Health Organization were directing their attention to the rampant diarrheal diseases of the children. However, Dubos observed a more basic problem and questioned whether the poor nutritional state of the mother while breastfeeding might be conditioning her infant's susceptibility to infection.[29]

As a result of these observations, experiments were devised in which a specific stress was introduced either to gestating or lactating mice or to newborn mice directly. To minimize other disturbances, the mouse strain, litter size, and diet were all controlled. The animals were kept under identical optimum conditions and followed throughout their lifespan. Since mice live more than two years, the task of feeding, caring for, and evaluating many animals took a large portion of research funds as well as enormous space. Russell Schaedler designed an ingenious housing complex of one-room apartment cages that were assembled into multistoried architectural structures, dubbed

Mousehattan.* This mouse community had a private cook, bottle washer, and housekeeper, thus allowing an efficient study of lifelong effects of mild disturbances that are experienced during the earliest days of life.

Early influences of infection and nutritional deficiency. These experiments produced some striking results. Any one of the mild stresses, applied a single time, produced irreversible lifelong effects, including adult size, weight, resistance to stress, longevity, even some aspects of behavior. For example, when the quantity or quality of protein was limited in the mother's diet during gestation or lactation, her young grew slowly, and this resulted in small adult size.[30] When newborn mice were infected with a virus taken from the intestinal tract of ordinary mice, it did not produce obvious disease, although it also resulted in underweight animals, and this trait passed on through several successive generations.[31]

The brief span of these experiments, from 1965 to 1971, precluded a systematic study of mechanisms through which the early influences were exerting their effects. Nonetheless, Dubos hypothesized that as a result of biological memory, infection and nutritional deficiencies could have interfered with one or more metabolic pathways to decrease the ability to incorporate amino acids, synthesize growth hormones or certain brain proteins, or develop muscle cells.[32] At this point in his waning research, he was satisfied to have elucidated a pattern of evidence showing how powerful environmental factors can be in physically molding the young and how they are more complex and subtle than had been previously studied.

There was no hesitation in taking this message public. In many lectures, Dubos warned that low birth weight is one of those little known biological dangers that carries a risk of lifelong physical and mental disabilities. Today, low birth weight serves as a major indicator for the United States Department of Health and Human Service's "Healthy People" initiative to measure the general state of health of Americans. A recent study found that rates of low birth weight are rising sharply, with suburbs outpacing cities; researchers attributed this to in vitro fertilization, which can produce multiple births, and technically advanced medicine to improve an infant's chance of survival.[33] Seriously overlooked, however, is the risk from increased use of antibiotics during labor and delivery to prevent infection in newborns, a chemical disturbance like malnutrition, infection, or toxins that the Dubos laboratory associated with low birth weight.[34]

*According to the "Talk of the Town" section of *The New Yorker* (4 May 1963), the laboratory playfully submitted the plans for Mousehattan to the New York City Department of Buildings with a request for a building permit. In the spirit of the request, the plans were denied, citing inadequate draining and plumbing systems and lack of fire escapes.

Early influences of crowding, pesticides, and pollutants. Danger from other early influences was revealed by the research of two graduate students in the laboratory. Alexander Kessler and Glenn Paulson explored the effects of early, gradual exposure to crowding or pesticides. These investigations, and the activities related to them, were important in expanding Dubos' early interest and involvement in the environmental movement.

Kessler studied a population explosion of mice in a unique facility designed to monitor their biological and behavioral processes. Over the course of a year, ten pairs of mice grew to nearly a thousand animals living in a density greater than ever before studied within a single enclosure. A surprise finding was that extreme crowding of adequately fed, watered, and otherwise clean animals was compatible with general physical health. Rather than suffering from obvious disease or increased adult mortality, the population endured unexpected problems related to self-regulating mechanisms, including high infant mortality, decreased social interactions, reduced fertility and fecundity in addition to aggressiveness, deviant sexual behavior, and cannibalism. The disrupted nurseries, lack of maternal care, and exposure to abnormally behaving adults were considered major influences on the subsequent adaptability of newborn mice.[35]

Dubos' view of crowding derived in part from this experiment and more strongly from experiences of living in and visiting many of the most crowded cities of the world. In contrast to Paul Ehrlich's "population bomb" and John B. Calhoun's "behavioral sink" that dwell on fears of death from disease and social deviancy due to population density, Dubos contended that human crowding exerts several independent effects, both benign and ominous. On the benign side, he argued that humans are naturally gregarious and even seek out all kinds of stimuli and contacts with people in crowded places, finding in them sources of great creativity and social evolution. On the ominous side, while he largely disagreed with those who claimed that crowding facilitates disease and conflict or who blamed the fall of populations primarily on the germ theory of disease, he saw disease more as a secondary phenomenon that followed metabolic disturbances. The real dangers of crowding, he concluded, were those silent biological injuries sustained by becoming progressively tolerant of filth, noise, ugliness, and visual confusion. Similarly, emotional injuries could result from adapting to the regimentation, loss of control, withdrawal, and severed links to natural rhythms spawned by crowding. Dubos blamed the distresses from (mal)adapting to crowding, rather than the crowds themselves, as the seriously overlooked culprit in producing diseases with no perceived origin.[36]

Paulson's research on early influences added evidence of harmful effects of *apparent* tolerance to environmental toxins. In his experiments, newborn mice were fed minute doses of the insecticides DDT and dieldrin at levels analogous to the low residues common in the human environment. Long-term exposure to the potent chemicals caused them to accumulate in fatty tissues of mice without causing disease; they also passed from mother to offspring. Symptoms of poisoning developed only after the animals were stressed, for example, when deprived of food. This stress triggered the release of large amounts of the chemicals stored in fat into the blood stream and led to disease or outright death. Rachel Carson had made a strong link between disease and exposure to DDT as it passes through the food chain. Paulson, however, showed that disease was more insidious and subtle, because the chemicals create unknown links and indirect responses.[37]

To Dubos, this study confirmed why looking for explicit carcinogens or mutagens had to go "beyond the description of a few *immediate* pathological responses. Hardly anything is known of the *delayed* effects of pollutants on human life even though they probably constitute the most important threats to health in the long run." His harsh view was that "the narrow-minded obsession with molecular biology, and the classical approach to pathogenesis through short-term experiments using one-variable systems [on acute effects], obviously will not provide the kind of theoretical knowledge needed to understand the indirect and long-range effects of drugs which disturb the physiological and ecological equilibrium of the whole organism."[38] Today's challenges from the unknown elements in genetically modified or irradiated foods, nutraceuticals, diet aids, and herbal remedies continue to reveal that an organic approach to toxicology is still not adequately dealt with or taught. An elemental truth persists in Dubos' motto: "There are no harmless chemicals, only harmless use of chemicals."[39]

Based on these experiments, Dubos publicly warned that not enough attention was being given to minute particles of lead in the environment. There was a "silent epidemic" poisoning children who ate paint chips containing lead that accumulated gradually, like DDT, in their bodies. As a result, he and Paulson organized a conference in 1969 on lead poisoning, sponsored by the Scientists' Institute for Public Information (SIPI), which had been founded in 1963 by Barry Commoner and Margaret Mead.[40] Experts at the conference identified the seemingly straightforward scientific and medical problems and described ways to detect and treat the condition before it caused irreversible neurological damage. *The New York Times* quoted Dubos' impatient rebuke that "the problem is so well defined, so neatly packaged with both the causes and cures known, that if we don't eliminate this social crime, our society deserves all the disasters that have been forecast for it."[41]

Dubos was unfortunately correct in expressing guarded optimism about ending this epidemic. The publicity generated by this conference led to regulating medical screening methods and punishing landlords for using lead-based paints. Nonetheless, more than four decades after these standards were put in place, children are still being poisoned first and treated afterward. Worse, it is now known that medicine to lower levels of lead in the blood does not reverse the neurological damage.[42] As Dubos had perceived about many medical problems, including lead poisoning, it was not enough to have knowledge, scientific expertise, and medical therapy in place to solve a problem. The more important goal is to create social resolve to prevent technologies not proven safe and if prevention fails to maintain active opposition to the practice until it is eliminated.

Another environmental pollutant came to the laboratory's attention when the Federal Trade Administration asked Dubos to perform experiments on bacterial enzymes from *Bacillus subtilis* that were being added to laundry detergents; these were causing severe skin and respiratory ailments in workers in the manufacturing plants as well as in consumers. Dubos, using his early expertise, found the bacterial enzymes were toxic to red and white blood cells, and his colleague Merrill Chase showed that low levels of exposure to them could sensitize individuals, aggravate allergies, and provoke subclinical infections.[43] Dubos' warnings against their use were picked up by science writers in many newspapers, by consumer advocate Ralph Nader on public television, and by investigative journalist Paul Brodeur for a *New Yorker* report, "The Enigmatic Enzyme." Dubos showed a rare anger toward the manufacturers for introducing these biological agents into detergents without testing and without considering the consequences, calling the act "criminally stupid."[44] The FTA's ruling on enzymes changed the way they were handled by workers but did not ban them from detergents. Whether consumer exposure to enzymes is hazardous remains an unanswered question.

Ideas for a Science of Human Nature

Concurrent with these bench experiments during the 1960s, Dubos was lecturing outside the Institute on social aspects of medical problems. Initially timid and worried about whether he would embarrass the Rockefeller community by speaking outside his professional competence, he wrote President Bronk to explain his numerous absences from campus. Bronk's reply encouraged him to pursue this double life in the laboratory and on the frontiers of a changing society and to speak out on "the need for more research on the nature of healthy man at a time when we spend so much for research on unhealthy men." Bronk also urged him to elaborate on the need for "greater cooperation between biologists and engineers in shaping our rapidly changing technological civilization so as to make possible more desirable ways of life for those who have, as you say, the capacity for adaptation."[45]

Dubos was not timid about devising various proposals for a novel, and to some an impractical, science devoted to human nature. He worked from the foreboding statement by French physician and humanist François Rabelais, that "science without conscience is but the ruin of the soul."[46] The science, as first envisioned in *The Torch of Life*, would "find it possible to consider within its province the study of man not merely as a machine or as one animal among so many others, but as a sensitive, imaginative, and ethical being who remembers the past and who lives emotionally in the future."[47]

The parameters of this science remained theoretical, postulating both a *scientific ideal* for which experimental knowledge would not reach the precision and predictability of most sciences and a *medical ideal* for which physicians would no longer act as "God's body patchers," to use Martin Luther's phrase, and hospitals would no longer operate as "repair shops." In his view, medicine, the oldest of sciences, was moving toward a confining body of techniques and knowledge to cure ills, retard death, and establish norms to which individuals must conform. Such medical "progress," he argued, would eventually cause an enormous increase in cruel dilemmas in medical ethics, involving decisions about which aspects of human nature to tamper with and which to respect at all costs. To avoid these dilemmas, or more wisely to prevent their appearance altogether, he envisioned a science to explore a broader biological knowledge of the human condition.

Trying to summarize Dubos' inquiries into this science is comparable to a comment by Hobart Lewis, editor of *Reader's Digest*. Lewis once introduced Dubos by saying he had spent his entire life digesting work of others but when it came to Dubos, he found it impossible to reduce the richness of his work to a few ideas or pages. In this new activity of defining a science, Dubos produced nearly a hundred essays and four books on a topic that preoccupied him intensely for a decade. While these efforts test ideas of traditional medicine, offer scholarly criticisms of science, and propose interdisciplinary strategies, they constitute a fragmentary but stimulating work-in-progress. Many major topics were tried under a variety of names and on a wide range of audiences: physicists, physicians, public health and government officials, architects, artists, and humanities professors. The three depicted below—medical sociology, wholistic biology, and humanistic biology—can best be appreciated by realizing they are all based on reintroducing health into medicine as ecological well-being. This concept cannot be found in an explicit passage from any essay or lecture; it is the leitmotif of the entire opus.

Medical sociology. In the same year Dubos wrote *The Torch of Life*, he also published a series of lectures called *The Dreams of Reason* (1961) where he introduced medical sociology as a science to serve mankind. Unorthodox, even revolutionary, this science would first have individual and social goals identified, something that would demand courageous choices based on desires, value judgments, and ethical standards, before research would be done to provide sufficient options for their solution. Some problems suggested for study included population growth, worthwhile employment for teenagers, medical technologies, and diets based on biological wisdom. This science would be a means to achieving goals set by society rather than an end in itself.[48]

As proposed, medical sociology would anticipate problems and report them. Anticipation meant, for instance, identifying ubiquitous and harmless microbes that could become pathogenic and predicting potential medical problems from seemingly innocuous environmental pollutants, malnutrition (over- as well as under-), radiation, automation, and compulsory leisure. Reporting would involve a community-based effort he called "prospective epidemiology," a global early warning system to report pathological trends from threats such as radiation, noxious chemicals, antibiotic resistance, and new vectors of disease in addition to a local alarm system in chemical industries for the millions of workers whom he described as "guinea pigs of the technological environment." Since he believed new diseases would also entail novel public health practices, any early warning system would surely aid in their prevention. Dubos would have enthusiastically endorsed ProMED, the program for monitoring emerging diseases, a daily Internet news report gathered from physicians, scientists, public health officials, and science journalists in more than 150 countries around the world.[49]

Wholistic biology. The disturbing fact that humans unconsciously adapt to almost anything led Dubos to develop the idea that inadequate responses to environmental forces created by modern life are "at the origin of most medical problems." Wholistic biology—the whole man in the total environment—is best developed in *Man Adapting*, where it is called "organismic and environmental" biology.[50]

As proposed, wholistic biology would acquire objective knowledge of adaptive responses of the whole man and would go beyond biological aspects to study social, cultural, and emotional adaptation. Some problems he suggested for study were conditioned reflexes, subconscious mental processes, sensory stimulation and deprivation, imprinting, and other behavioral phenomena involved in adapting to new environments and challenges. This would be similar to research done to train soldiers for combat in the tropics or the arctic and astronauts for space travel. Consequently, knowledge gained from this kind of biology would enable humans to manipulate their surroundings sensibly and to cultivate biological mechanisms on which to build healthy, creative lives.

With its emphasis on human-based knowledge, wholistic biology was at the forefront of an ecological attitude toward scientific medicine, one that continues to develop. At the time, however, Dubos worried that his ecological approach would find more favor among sociologists than medical scientists. Defending his approach to Cornelius van Niel, whose famous course in microbiology included the ideas of Dubos, he wrote, "organismic concepts are being lost from sight and replaced by a most unsophisticated reductionist philosophy. It is for this reason that I take any opportunity that presents itself to plead in favor of organismic and ecologic biology and medicine."[51] As he wrote to the puzzled molecular biologist who was translating *Man Adapting* into Japanese, "I wanted only to convey the view that there are larger aspects of biological sciences, other than molecular biology, that are important for human health . . . molecular biology and environmental (organismic) biology are the two complementary aspects."[52]

Man Adapting, in fact, reached a receptive audience among physicians who were eager to think about patients in the larger context of their daily lives. The book served as a text in medical schools for three decades and influenced the creation of many academic departments and not a few courses devoted to environmental medicine. According to David Mechanic, the René Dubos Professor of Social Medicine at Rutgers University, "More research than ever before takes place at the boundaries of traditional disciplines . . . [providing] an interdisciplinary paradigm increasingly useful to both investigators and clinicians."[53]

Humanistic biology. The idea of humanistic biology was introduced in 1965 as a science to search for additional "biological truths" that account for and add greater meaning to humanness.[54] The idea took shape after completing *Man Adapting*, during a sabbatical at the Center for Advanced Studies of Wesleyan University in Connecticut, seminars at liberal arts colleges, and conversations with many inspiring humanists who were invited to Rockefeller for sabbatical visits, including Loren Eiseley, Marjorie Nicolson, Joseph Wood Krutch, and Archibald MacLeish.

At first glance this oxymoron, humanistic biology, appeared artificial and abstruse, yet the concept was surprisingly well received by those who were disenchanted about the relevance of science to human life. Chemist, author, and *Encyclopedia Britannica* editor Louis Vaczek praised Dubos for producing a new biological "portrait of Everyman" and "seeing new and more comprehensive structures beneath familiar surfaces, relating them all to the history and the presence of the individual."[55]

Also called a "science of humanity," and a "science of man and humanism," humanistic biology invited exploring biological aspects of human life that are largely ignored by reductionist scientists. Two successful examples of developing "objective methods for describing all aspects of reality" from the recent past, cited by Dubos, were the doctrines of evolution and the great chain of being that provided biological evidence relating man to the rest of creation. Whatever questions are asked about human nature, he said, "the answers of theologians and of philosophers must be consistent with the demands of informed intelligence—that is, with scientific facts."[56]

Research in this science would identify "patterns of responses" to the environment that are organically conditioned by past experiences, social structures, emotional attitudes, and ethical concepts—the *formative* responses that lead to creating various ways of life. More precisely, he might have called these "patterns of health" to balance them with his earlier concept of "patterns of disease," or *failed* responses to environmental stimuli. The purpose of these studies would neither account for individuality nor define an ideal human being but instead "provide knowledge of the raw materials of experience out of which man creates himself." During his final years, Dubos led the way in identifying a few patterns of response that he decided had not yet been recognized by science, let alone the humanities (discussed in chapter 6).

Some audiences questioned whether humanistic biology was meant to bridge "the two cultures" as defined by C.P. Snow.[57] As mentioned earlier, Dubos never accepted the humanities and sciences as separate cultures lacking common knowledge and unintelligible to one another. Rather, he said, "There are a multiplicity of intellectual occupations, each of which fortunately has several points of contact with human life." Whatever his specialization, the scholar must "raise his language above the jargon of his trade . . . [and] learn to speak to man." In rare mentions of Snow's lecture, he chided him for updating his views of the arcane scientific details that a layman needed to know—supplanting thermodynamics and entropy in the essay's first edition with DNA structure in the second and with ecological principles and general systems law in the third. He took advantage of Snow's behavior to convey how fashions change enormously in science, and how scientific knowledge is never absolute or final.[58]

Countering Snow's view that biology had lost contact with the humanities because it had become too scientific, Dubos judged that "biology is not scientific enough" because it was cultivating abstract aspects of life with such intensity that "its perception of nature in the round [had] become atrophied, or at least blunted." What was needed was "an organismic and ecologic attitude very different from the analytic one which now prevails in biology."[59] While expecting humanistic biology to generate solid facts and exacting laws, he did not presume its scientific knowledge would be grander than what emerges in the humanities from abstract thoughts or vivid imagination. Its greatest potential would be to sharpen and extend the direct perceptions from which art, literature, and philosophy have always emerged. "Reality has multiple facets," he observed, "and, therefore, can be apprehended only if seen from different points of view."[60]

During the 1960s, advancing a science of human nature was, at best, provocative, diffuse, and provisional. Dubos fully expected that some hypotheses would be proven wrong or impractical. After all, he was writing against an orthodoxy rooted in a scientific revolution begun by René Descartes three centuries earlier. Some criticisms about whether higher degrees of integration and interaction could be studied experimentally came from chemist Melvin Calvin, who maintained that science demands simple empirical parameters, and physicist Henry Margenau, who argued that science was not designed to generate values.[61] In addition, humanist Jacques Barzun thought Dubos was confusing influence and conditioning with manipulation and he questioned whether something as abstract as a pattern of response could determine practical or intellectual action.[62] Dubos acknowledged that his view of this science was uncommon and would need an uncommon approach. This would require a new institution—perhaps one like The Rockefeller Institute that was formed when medicine was emerging as a science—one that would be independent of existing institutions and able to "will new departures."

A clue about how he saw his role in this activity came in a seminal lecture before the American Philosophical Society on "Logic and Choices in Science."[63] Before a largely scientific and potentially hostile audience, he declared that the logic and orderliness produced by the experimental method were an "illusion, or the result of an afterthought" whose scientists help science move along its predictable course. More realistically, he argued, science continuously changes directions, and its innovations "reflect the attitudes and concerns of the social environment in which they arise." Therefore, scientists are also needed, and he included himself among them, to warn of problems to come, to anticipate solutions, and "thus open new channels for the human mind." His vision for a new direction for biology—a science of human nature—was to learn how humans respond to the total environment and thereby "create living as experience." The lecture's rather humble conclusion is that studying phenomena of the living experience was "often a pathetic effort, but one that enlarges human vision." Lewis Mumford later wrote to Dubos, acknowledging how this lecture with its "humanizing of the scientific world picture" had given him confidence to complete *The Myth of the Machine.*[64]

During these years, Dubos also tried putting into practice what he was advocating. Not all these experiences in promoting a science of human nature were successful, although they yielded enough feedback to modify his thinking in the ensuing decade.

One opportunity to put these ideas into practice came when he was asked to be the first director of a new organization associated with Columbia University. His physician-scientist friends and Nobel laureates in medicine, André Cournand and Dickinson Richards, founded The Institute for the Study of Science in Human Affairs in 1966. Its original mission was an outgrowth of the French intellectual movement Prospective, created by French philosopher Gaston Berger in the 1950s, which debated how humans could take a constructive role in creating a desirable future; this appealed to Dubos. Then, for unknown reasons, the mission was recast to emphasize the need "to correct the social absurdities and monstrosities that result from the mismanagement of scientific technology," which caused Dubos to refuse the directorship, formally noting these goals were outside his competence in technology. His more explicit objection, with suggestions for a more socially relevant mission, appeared in his 1967 inaugural lectures for this institute that were published as *Reason Awake: Science for Man* (1970). Here, revisiting his ideas for medical sociology, he maintained that scientists and academics alone cannot do the job. Science may provide a factual basis for options, he argued, but an informed public must consider consequences and risks in making decisions. The scientific study of social problems "requires a complex, integrated approach not readily achieved within the present academic structure." Nonetheless, the difficult task of integrating this institute into a traditional university structure failed and it dissolved within a few years.[65]

Human Ecology: An Interruption

Dubos is often labeled a human ecologist. To a certain extent this designation is correct, but only in the philosophical sense of a lecture he delivered in 1969. For a few years, Dubos seemed to think his science of human nature might lead in the same direction as theoreticians trying to develop a hard science of biology called human ecology; by 1972, he abandoned the term.

While many concepts of human ecology reach back to ancient times, its history is not much older than the twentieth century. The first use of the word ecology, from the Greek *oikos* for house or habitation, is credited to German biologist Ernst Haeckel in the 1860s to describe the web of relationships among organisms and their communities.[66] Ecology did not become an active field of study until after 1920 with the founding of societies and journals devoted to the subject. As noted earlier, one of Dubos' first publications appeared in the nine-year-old journal *Ecology* in 1928, where he reported environmental effects on decomposing cellulose. Also during the 1920s, geographers and sociologists were among the first to study human ecology at universities of the American Midwest, and they attempted to adapt current concepts of biology to study human systems; their scholarly history, however, reveals they considered human ecology a basic social science.[67]

During the 1930s, some influential writers called for making human ecology a hard science. Social philosopher Lewis Mumford in *The Brown Decades* (1931) thought human ecology should provide "a dynamic interpretation of basic social and economic relationships."[68] Historian, novelist, and learned advocate of science, H.G. Wells tried to mandate an "ecology of the human species" in his trilogy linking perspectives of history, biology, and economics. He defined ecology in *The Science of Life* (1935) as the science of "vital balances and interchanges" in human life that "might have been a better and brighter science if it had begun biologically."[69] Botanist Paul B. Sears in *This is Our World* (1937) began to conceive an ecology of man that studied the interactions of cultural patterns within physical, biological, and social environments.[70]

Twenty years later, in 1957, Sears was continuing to fashion an integrative hard science of human ecology and urged rewriting history and restudying human values, "with an eye to Man's long evolutionary background and his growing role as a natural force. What environment does to him and what he in turn does to it is of far more significance than the loves of monarchs and the quirks of generals." Ten years later, Paul Shepard was asking in *BioScience*, "Whatever happened to human ecology?"[71]

All the while, orthodox ecologists were limiting their studies to the effects of human actions on other species and on the environment. Dubos, in tacit agreement with Sears, Shepard, and a few other theoreticians, argued that what was needed was a science focused on human beings. The problem for these ecologists was not whether man would or would not transform nature, but how he would and should do it. This view was perceived as anthropocentric, especially by environmental historians such as Roderick Nash as well as by wilderness and land preservation activists.[72]

By his own admission, Dubos practiced "enlightened anthropocentrism," in which what is really good for man is also good for earth.[73] Like Sears and Shepard, he carefully distanced himself from pure anthropocentrism where the human species is the only value to be considered in managing the world. Enlightened anthropocentrics believe humans manipulate the earth for their best interests *only* when they first love earth for its own sake. "Even when man attempts to look at the external world objectively," Dubos wrote, "he still sees it with human eyes, interprets it with a human mind, and evaluates it on the basis of human assumptions and values."[74] By studying humans within natural systems, a true human ecology would make it possible to understand how their activities could nurture such values as diversity, stability, and beauty and promote health for themselves and the earth.

Other terms, more recent than anthropocentric, including biocentric, ecocentric, or deep ecology, do not describe Dubos' views either.[75] He focused on individuals, their quality of life, and experiences within their own surroundings; he avoided the sweeping relationships of populations with their natural environment (geography, landscapes, ecosystems) studied by orthodox ecologists. Even with the larger problem of environments in mind, the experimenter still had to be pragmatic and work on an available, individual aspect of it. "Human ecology," he said, "deals not with man in the abstract, but with particular men at a particular time in one particular place."[76] Dubos was in complete agreement with Shepard's statement that "There is only one ecology, not a human ecology on one hand and another for the subhuman . . . it means seeing the world mosaic from the human vantage without being man-fanatic."[77]

During 1969, Dubos continued to grapple with definitions of human ecology and speculated whether it should instead become "a philosophy based on a synthesis of scientific knowledge and social conscience." In the 1969 Jacques Parisot lecture titled "Human Ecology," sponsored by the World Health Organization, he took the opportunity to realign human ecology with positive, beneficial effects of the environment, partly as a reaction to the word *environment* that he complained was being used only to evoke crises, degradation, and depleted resources.[78] Then, a few months later in the British *Science Journal*, he also disparaged *human ecology* for being "almost exclusively identified with the social and biological dangers that confront man. Yet there is much more than this one-sided view of man's relation to his environment."[79]

A year later, in *Reason Awake: Science for Man*, Dubos jettisoned all efforts in promoting the term, stating, "Human ecology is a no man's land" and admitting that "we lack methods for investigating scientifically the interrelatedness of things."[80] Sears' review of this book reiterated this opinion that findings in ecology do not lend themselves to experimental control. "Whatever benefits may come from ecology," Sears wrote, "will have to be expressed by changes in human values and behavior."[81]

To a large extent, what prevails today as human ecology is ecological thinking, or "philosophy" as proposed in the Parisot lecture. This is the kind of thinking that characterized Dubos' lifelong approach to problems and what Sears admired as the "clarity of the French mind at its best." Even in 1983, biologist Gerald Young noted that human ecology was still not firmly established "within any one discipline, even in the social sciences, and is perched rather precariously on the margins of each and every academic realm involved with the study of the human species."[82] Akin to what Dubos envisioned for a science of human nature, biology contributes only one aspect of an interdisciplinary approach to explore complex problems in their context, whatever their origin and nature. Today, human ecology is the focus of many courses, degree programs, academic institutions, and an international society founded in 1979 with a mission to reach "a fuller understanding of the place of humans in nature."[83]

Remarkably, human ecology is not even mentioned in *So Human an Animal*, the book written during the very years he was trying out this term.

Taking Another Direction

So Human an Animal emerged after Dubos received the 1966 Arches of Science Award from the Pacific Science Center in Seattle, Washington. The award, headlined on the front page of *The New York Times* as the "American Nobel," was for "conveying the meaning of science to contemporary man." Dubos told the newspaper's reporter that the cash prize would go toward trying "to save the Hudson River Valley" by transplanting and protecting another thousand hemlocks.[84] The prize did allow the Duboses to replace their aging Studebaker and to move from a tiny rent-controlled apartment overlooking the main entrance to Rockefeller, where they had lived for twenty years, into a penthouse apartment a few blocks away. More important, the prize brought his work to the attention of Kenneth Heuer, an editor at Charles Scribner's Sons, who had published, to wide acclaim, several philosophical and somewhat mystical books by anthropologist Loren Eiseley on man's place in the universe. Heuer encouraged Dubos to leave behind the science experiments and contemplate what he believed should be the proper study of mankind in a world of technology.

Dubos enthusiastically transformed *Man Adapting* into *So Human an Animal: How We are Shaped by Surroundings and Events*. Leaving out the science and engaging the world with common sense and plain language changed a book for experts into a literary work of universal interest. The basic theme was retained, that man has a remarkable elasticity and can adapt to any demands of the environment, for better or for worse. In contrast with *Man Adapting*, which interpreted disease as failed adaptations on a medical level, *So Human an Animal* interpreted health as potential readiness to anticipate problems and to respond with "creative adaptations" to lifestyles and experiences.

So Human an Animal was awarded the Pulitzer Prize in 1969 for its provocative observations and timely warnings about environmental dangers and their effects on human health. The threats he identified and the alarms he raised, coming from a widely respected biologist, garnered headline stories. Audiences responded quickly to Dubos' fondness for hyperbolic charges such as we would be "remembered as the generation that put a man on the moon while standing knee-deep in garbage."[85]

The book balances the threats and alarms with constructive views that something could be done to restore the quality of life. The final chapter, "The Science of Humanity," argues less rigorously for a specialized science of human nature than he was doing elsewhere, and more vigorously for a new ecological attitude "so unfamiliar, even to many scientists," which would plan for the future by developing human potentialities and pursuing the significance of life, not its mastery.[86]

In 1971, the laboratory of Environmental Biomedicine was disbanded when Dubos reached the University's mandatory retirement age. As an experimenter, he acknowledged that his limited search for environmental determinants of disease was "not a fashionable topic and carries little scientific prestige."[87] Nonetheless, during this final decade of experimentation, he operated from the conviction that it represented "scientific capital," or the identification of problems being neglected because their solution was not in sight. Future explorations, he realized himself, would require entirely different knowledge, questions, techniques, and even institutions to support the work. The extent to which environmental medicine becomes a valid discipline may take decades of laboratory science, even after much useful knowledge has been gained from molecular biology and genetics. Significantly, however, this research served as a base for his biological philosophy of human and environmental health that he professed for another decade.

Amid polarized views over the relation of scientific medicine to human well-being, Dubos stood between those who believed science and technology would conquer disease and others who believed they would not, despite expensive technologies. Like his friend, physician-scientist Lewis Thomas, Dubos spoke openly about why there could be no technological guarantees of health and how medicine would lose its unique position among the sciences unless it accepted some responsibility for the various aspects of life that determine our humanness. Thomas once remarked how Dubos had "learned more about medicine than most physicians" and knew "that mankind's changing of his own environment has had much more to do with susceptibility or resistance to infection than anything in the modern pharmacopoeia."[88]

To that end, Dubos actively disputed the desirability of conquering any part of nature and remained optimistic that a new science would someday emerge to define the natural place of humans among microbes and sundry other toxins and physiological stresses. By understanding adaptations, both those of our adversaries and our own, and by designing novel strategies to restore the natural checks and balances in disturbed relationships, he believed, in the words of French Revolution pamphleteer François Lanthenas, that "Medicine will be what it must be, the knowledge of natural and social man."[89]

As a result of the more successful negative arguments in *So Human an Animal*, Dubos found himself in a popular but awkward position, for him, of being admired as a doomsayer, and he became perturbed that his ideas to promote connections between environment and health had been lost.

Impatient, or restless once again, he took up a distinct, robust position that was neither pure environmentalist, conservationist, preservationist, nor any other special defender of earth. If humans are blamed for all the thousand devils of the environmental crises, he said, then humans can also prevent and solve them. Unlike activists who fought to save environments, Dubos championed the conservation of mankind.

During the next, and final, decade of his life, Dubos emphasized positive relationships between humans and their environments. In place of trying to fashion a new science, this biologist turned to giving voice and structure to what he called "the shapeless aspirations and preoccupations of the multitudes."[90] Just as he had disturbed orthodoxy in medicine and in science, he set out to disturb complacency about "being human."

6 Health as Creative Adaptation

All thinking worthy of the name now must be ecological.
LEWIS MUMFORD

FROM THE NORTH-FACING WINDOWS OF HIS OFFICE in the Bronk Laboratory of The Rockefeller University, the elderly Dubos would become lost in thought as he gazed down on the sanctuary that framed his life as a scientist. At treetop level, above the well-tended gardens, he could take in the whole landscape. On the right were the Institute's beige brick buildings. Nearest was the Hospital, where Oswald Avery welcomed him on the sixth floor in 1927, and farthest was Smith Hall, where he ran his laboratory from 1944 until 1960 when he moved into the new Bronk Laboratory. Looking three blocks straight ahead, just beyond the wrought iron fence surrounding the Rockefeller campus, his eyes met the towering New York Hospital. This white gothic skyscraper, which reminded him of the Palace of the Popes in Avignon, was the place where he lectured regularly to Cornell University medical students, where his wife Marie Louise died, where he several times had been a grateful patient, and where he would later die. On the left were the modern limestone and spandrel glass structures of the University that served as student residences, lecture hall, and administrative offices. To him, the architectural simplicity and utilitarian nature of these low-profile buildings reinforced their sturdy purpose as workshops for research and shelters for scholarly thought. In his biography of Avery, Dubos praised the freedom he also enjoyed in this cloistered space with its "priesthood" of scientists, its walnut-paneled common rooms, and its long *allées* of sycamore trees arched over lawns, gardens, and marble walks. The atmosphere of this venerable institution sustained and enriched his contemplative nature for more than fifty years.

Turning away from the windows and into the fourth-floor room where he thought and wrote, Dubos sat in a spare, muted, and nearly unadorned space. There was no direct approach to this room from the main hall. Entry was from either his private laboratory or an outer office. Once inside, a visitor noticed a few shelves with books, a set of encyclopedias, and science journals along each side of a massive blond oak desk. However, most visitors were so intent on their encounter that they rarely noticed two very personal objects in the room.

Both objects were souvenir pictures from his French youth, icons of the robust temperament he had continued to cultivate since those days. Between the two windows and behind the desk was a small oil painting of Don Quixote tilting against a windmill. This copy, based on a Daumier painting, was made by Auguste Panon, who had taught him how to paint. Visitors to the office faced this knight on horseback while they engaged Dubos in conversation across the desk. For a profile in *The New York Times Magazine*, reporter John Culhane asked about this painting. Dubos lightheartedly commented, "I hang it here as a symbol of the fact that we scientists are often tilting at windmills—problems that we invent rather than problems of reality." Inspired by this exchange, a photograph of Dubos with the painting appeared in the profile that was aptly titled, "*En Garde*, Pessimists! Enter René Dubos."[1]

The other picture in the room, an oversize photograph of a gargoyle on Notre-Dame of Paris, confronted Dubos every time he looked up from reading or writing. Called by some the spitting gargoyle, the figure is more popularly known as *Le Penseur*, the thinker. With head in hands, this horned and winged creature, who is part human, angel, and devil, sticks out its tongue and glowers with amused bewilderment on the antics of human life in the city below. It stands at a strategic corner of a balustrade looking, ironically, like it is both fleeing from the cathedral and guarding its outer walls and towers. To Dubos, as to medieval Christians, the gargoyle was more than a fanciful ornament; it symbolized an accepted coexistence of the profane with the sacred. Its human size also provides scale to the grand monument on which it perches and serves to keep the world view within human contexts and perspectives.

The two lone figures in these pictures represented to Dubos something of life's mysteries, of imagination within tradition, and of iconoclasm amid orthodoxy. Their presence on his office walls quietly mirrored his own defiant outlook on the world. During his lifetime, Dubos had acquired a certain *joie de vivre* of the knight-errant by acting on his own dreams with combativeness, adventure, and risk taking, qualities he proclaimed in *Mirage of Health* that were essential to life. Likewise, he consciously developed daring views, reminiscent of the gargoyle's profane stance, and was frequently astonished and gratified that his peers and public audiences expected and even sought out these views.

Within this space, Dubos could be intensely pensive and spurn the world as he sat alone and thought. In fact, the outward manifestations of his life were plain and simple, sheltering an intense dedication to his work. His work days had a timeless quality. Every morning just before nine, he walked the few blocks from his apartment to the University, always attired in an elegant suit, tie, and polished shoes. While fastidious about clothes—his suits were sewn by a tailor on Fifty-seventh Street—he avoided ostentatious display, except on St. Patrick's Day, by keeping new suits in a closet for several years before wearing them, so if asked, he could modestly reply that it was one he had for a while. On reaching his office, he replaced his suit jacket with a tan laboratory coat. On St. Patrick's Day, his self-described favorite holiday, he sported a special green linen vest under the coat.

During the active years of bench research he went directly to the laboratory, eager to record overnight results, review protocols with colleagues and technicians, and plan the day's experiments. After retirement and without the pressure of doing experiments, he simply sat down at his desk and began to revise and recopy the innumerable pages written the preceding night. Every manuscript went through many handwritten drafts. While concentrating, he absentmindedly twirled wisps of hair on top of his head, almost as if stirring up ideas inside.

There were occasional breaks from reading and writing throughout the day when he engaged laboratory colleagues in long, involved conversations. These exchanges played a deliberate role in his work and were opportunities to reformulate ideas he was struggling with on paper. These were also times for indulging in hypotheses, some intentionally extravagant, based on his eclectic reading. Many discussions carried over to the doctors' Welch Hall lunch room, where sharing a table with him was an unpredictable experience. Dubos often praised this space as the "the greatest educational system I have ever known anywhere. I came to the Institute not knowing a word about medicine. But every day in the dining room at lunch I became slowly sensitized. . . . My suspicion is that if it had not been for the dining room at the Rockefeller I would not have been as rapidly successful in science."[2] He claimed with some exaggeration that "there never was a symposium—in the etymological sense of the word, namely, a convivial meeting for drinking—that was more scientifically productive and intellectually pleasurable than those held daily in the lunch room . . . though coffee and ideas were the only intoxicants."[3] Where he once sat at tables dominated by Thomas Rivers and talk of viruses or by Alfred Cohn and lessons in the history of medicine, he now led discussions on the latest scientific news or on a profusion of worldly curiosities in history, art, and literature. Whether the topic was growing truffles in the laboratory, infections that produce beautiful variegated tulips, or why water does not freeze in fire hydrants in winter, he enlivened every meal. When the lunch room was replaced by a cafeteria open to all, he walked home to eat with Jean.

His personal life also became simpler as he focused on writing and lecturing. He did not indulge in sports, hobbies, or clubs, with one exception. From 1948 on, he relished his membership in The Practitioners' Society, and its monthly dinner-lectures at the Century Club, where he was in the exclusive and stimulating company of New York City's preeminent physicians.

The Duboses did not take conventional vacations, preferring to arrange family reunions or brief visits to the French countryside in between European speaking engagements. Most commonly, he and Jean selected a village that was "not a center of tourism where we can walk a little" especially medieval villages in the Dordogne they had read about in the magazine of the Touring Club de France.[4]

Their last apartment, reflecting his penchant for solitude and calm, was on the top floor of a cooperative near Rockefeller with commanding views over Manhattan. In his words, the interior had a "monastic character." The huge living room, with a polished wood parquet floor, had three heavy antique European chests against white walls and four matching chairs on a small rug. The very few objects in the room evoked personal associations, particularly a small ceramic Virgin holding the Christ child and a wool folk tapestry they designed that depicted a mixed urban and natural landscape seen through a monastery's triple-arched windows. Dubos told the weaver, after it was hung, that the tapestry added "a rich atmosphere of country jolliness." His bedroom, where he preferred to read and write into the early morning hours, included an old "cowboy" bed and Navajo blanket that he got in Arizona while doing field work at Many Farms, a wall of books, and a black leather arm chair.

Most evenings, he read avidly in science, medicine, and history, usually absorbing two or more books a night. His modest library was supplemented by books borrowed from the Rockefeller library, where he went daily to peruse the prominent displays of new journals and books. His method of deciding whether to take a book home was to read the first and last sentences to see whether the author's concluding statement was justified by the opening one. For those authors who met his criteria or who intrigued him, he proceeded to study the entire book, taking a few notes in longhand or memorizing images and examples in much the same way he had committed schoolroom *dictées* to memory.

For pleasure, he read and reread the journals of Thoreau and twentieth-century authors Albert Camus, Jean Giraudoux, and André Malraux. He was little interested in newspapers or current events, observing he could get the essential news from the radio while doing something else, although in the late 1970s, they bought a television and enjoyed the *MacNeil-Lehrer Report* and an occasional symphony or play on public television stations. The more he watched, the more he worried about the noise and chatter and about so many people who "had lost their taste and pleasure of experiencing the real world with their senses. . . . How can anyone know spring by seeing a tree in flower on television?"[5]

At this age, his life was settled, he was surrounded by congenial colleagues, he was economically secure, and both his and Jean's health were comparatively stable. Like Avery, the choice of a reclusive lifestyle, no matter where he lived, worked, and traveled, meant he deliberately avoided all kinds of busy-ness or socializing that most people assume is necessary for success. Dubos enjoyed the countless hours thinking about what really important things needed to be done.

A Visible Ecologist

In strong contrast with these contemplative work days, Dubos' vibrant public personality as a lecturer could be quite startling. Whether in front of a scientific, literary, government, business, or international audience, his performances often left an indelible (albeit mistaken) impression that he lived the glamorous, even swashbuckling, life of an activist.

Several prominent scientists became visible environmentalists during the 1970s, among them Margaret Mead, Linus Pauling, Paul Ehrlich, Barry Commoner, and Dubos. This was a time when scientists were beginning to come out from behind their laboratory benches, partly in response to a growing attitude that scientists were losing the confidence of the general public by not using their expertise on technical aspects of environmental issues. "A visible scientist" as defined by Rae Goodell in her book by that title, was simply one with "visibility to the general public," but her criteria did not necessarily include an influential or a public interest scientist. These individuals were attractive to the press for their hot or controversial topics, maverick or iconoclastic status, colorful images, and credible reputation. In her estimation, Dubos was one of the top fifteen of the nearly fifty visible scientists she interviewed, but he was not as highly visible because he was much older, with his "grandfatherly charm and an expressive French accent which is reminiscent of Maurice Chevalier," and less controversial politically than the others.[6]

How did Dubos compare himself to these visible scientists who often joined him on lecture platforms and programs? Paraphrasing comments from several interviews and lectures, he praised all of them for being more dynamic, articulate, and effective speakers because they created excitement and got audiences involved in details of scientific problems. In contrast, he thought his success had more to do with his French origin and education that distanced him from these down-to-earth American scientists. He quoted his French professor of sociology who taught, "Messieurs, here is what one sees. But here are things one cannot see." From him, Dubos had learned in writing and speaking "to be aware of what one does not see, instead of being satisfied with what one sees."[7] This lesson further suggests why he spoke in a questioning, even conversational, manner rather than a dogmatic one. His success, he added, might also be attributed to talking on a different level about human concerns and how to understand and deal with the science problems rather than about problems themselves. Lastly, Dubos observed, he was less controversial because he avoided discussing political issues in lectures, frequently telling audiences and journalists alike, "I have an absolute rule that, not having been born in this country, I never get involved in political matters."[8] Nonetheless, these differences did not keep him from being provocative.

Far more arresting than the appealing nature of his public performances, as he noted, was the newness of his ideas and topics. With an eloquent flow of stories and intelligence, he moved audiences to laughter, anger, tears, and routine standing ovations. Sometimes he sounded like an evangelist spreading news, masterfully implying, without ranting, that if these concerns were as important to them as they were to him, they should also get involved. There were constant, gentle reminders of collective responsibility for offenses against the environment and for finding "several safe roads to ecological salvation."[9] These biologically based "sermons" explored human dimensions of how "we must learn," "we must identify," "we must limit," or "we must change our ways of life" by developing positive values.

Given his usual inwardness and quiet life, Dubos was not prepared for the widespread response to winning a Pulitzer Prize for *So Human an Animal.* Its ensuing publicity and attention invaded his privacy and tested his emotional reserve. While he had long maintained a vigorous lecture schedule at a prudent pace, his audiences changed suddenly and substantially after the Prize. During the period from 1960 to 1981, he gave nearly nine hundred public lectures, an average of forty a year or almost one a week, an agenda that diminished only slightly by 1981.[10] About one-third of these took place in New York City and at least three a year were outside the United States. Before the Prize, more than half were academic lectures to medical and scientific audiences with the balance to college science students. After the Prize, only a handful were before these groups each year.

The new audiences included students and faculties, environmental organizations, professional societies, and local community groups.* In 1970, the year after the Prize, he gave seventy-five public lectures, the most in any single year. With increased visibility, he reached an enormous number of uncommitted students, committed activists, and concerned citizens, the very people he believed could put into action what he professed in words. He liked to tell the story, eyes twinkling, that when someone asked him to speak, he always suggested three titles "to make them feel good." What they did not know was that a topic for that audience had already been decided so the lecture would be the same regardless of the title they chose. It is not uncommon today to find someone who heard him lecture say that they had never forgotten it. For every lecture, a manuscript was carefully prepared, although he spoke from it rather than reading it, so the published lectures lost the spontaneity of asides and local examples he added while speaking.[11]

Before the Prize, Dubos had shunned publicity, respecting the University's practice of keeping a low profile in the news and broadcasting media. When the Prize was announced, Rockefeller did not have a public affairs office, so the thunderous attentions showered directly onto him. His timid reaction was to stop answering the telephone for fear someone would extract another interview or lecture commitment. Nonetheless, within a few months, he became a popular guest on radio and television programs nationwide, and it goes without saying that he enjoyed this attention, for a while. Soon enough, it became impossible to say no to many demands, and these began to complicate his life enormously. To regain some control, he accepted advice from a

*After retirement, Dubos accepted appointments to teach semester-long courses at three colleges. He initially relished interacting with undergraduates interested in environmental studies, but this enthusiasm was quickly dashed. His courses, outside the conventional curricula, dealt with problems he was confronting in his current work. Bard College colleague Michael Rosenthal observed in the *Clearwater Navigator* that "he had not thought it necessary to develop the techniques by which we professional front-line teachers push the distracted intellects of 18–21 year year olds to the frontiers of knowledge. In spite of his educational naivete, his students sense[d] something very special there, and . . . worked hard for him, and they learned." However, disappointed with sweeping generalities in the students' papers and discussions, he observed they reflected good training in humanities but not in natural sciences. Being a generalist, he knew, was not enough for effective action. His ensuing advice to young people was "Don't get involved in vague problems. Begin with a substantial and practical activity, and show you can accomplish something. After that the world will come to you." He resigned from college teaching, admitting "that if I'd continued to teach, my influence would have been a bad one—no a dangerous one."

public relations consultant and allowed a lecture agency to manage his speaking commitments. Within a year, Dubos ended this arrangement, exhausted by haphazard travels and disappointed that audiences who paid fees were not those he wanted to reach with his messages.

Unfortunately, there were situations where he found himself unsuited, awkward, even vulnerable. Among the particularly difficult commitments were those for which people sought his advice on technical and political issues, despite his protests these were beyond his expertise.[12] His resolve to shun politics disturbed those who sought his visible participation. Gerard Piel, long-time editor of *Scientific American*, noted how "he had to stand down politely from the nomination to cult figure that some young environmental absolutists pressed upon him."[13] Likewise, Dubos made no statements, private or public, on such major controversies as nuclear test ban treaties, abortion, or the Vietnamese war.

Nonetheless the press and media constantly sought him out for his independent, optimistic views. One way of maintaining a subdued profile yet reaching influential audiences were several op-ed pieces in the *The New York Times*. During the energy crises of the 1970s, Dubos joined debates over scarce or limited resources and social priorities with editorials titled "Human Life Can Prosper with Spartan Ways," "Less Energy, Better Life," "Creating Farmland Through Science."

On scientific and technological matters, he offered social rather than technical comments. Whether the topic was plutonium in nuclear power plants, destroying nuclear weapons, or regulating recombinant DNA, his position was similarly resolute: destroying or banning these technologies and their stockpiles would not bring peace or stop progress, for the simple reason that the knowledge of their manufacture and use can never be eradicated. Today's controversy over stem cell research would have merited this same rebuttal. Dubos reluctantly supported a nonbreeder reactor for nuclear power because of significant human and environmental health hazards from coal-based energy, yet he had misgivings about nuclear power, not because of its radioactivity but because it would increase social complexity and centralization of the power structure. He favored several alternatives to nuclear power and promoted using less energy from fossil fuels and more human energy; producing energy from renewable sources such as plants, wind, and thermal pools; and inventing unique regional solutions to energy needs.[14]

Along the way, there were also two amusing misadventures. In one, Dubos supported author Norman Mailer's 1969 campaign for mayor of New York City as an informal environmental "advisor." These two, who shared the 1969 Pulitzer Prize for Nonfiction, came together on a highly unusual platform that New York City should secede from New York State. Despite this quixotic goal, the counterculture antics of this curious twosome stand out in the annals of failed mayoralty races. Another gamble was joining the National Campaign to Make General Motors Responsible. This group, headed by Ralph Nader, charged GM with producing automobiles that generated 35 percent of the nation's air pollution, lacked safety features, and were designed to incur expensive repairs. Dubos was nominated in a nationally publicized campaign to serve as one of three GM board members representing the public interest.[15]

These two ventures collided with the arrival of a new president and student storms of protest at Rockefeller. While converting the Institute to University, former President Bronk had added buildings, extended academic departments, and embarked on a lavish Ph.D. program. The new president, Frederick Seitz, former head of the National Academy of Sciences, was confronted with restoring financial realities, and his decisions were not being well received by the faculty and students. When Dubos brought Mailer to the University for a much publicized campaign speech, and when he joined Nader's group, these maverick actions helped kindle the unrest. In April 1970, Dubos wrote Seitz, "I have no feeling of hostility toward GM (the Buick car I own has served me well). I have joined the Nader movement, because I believe that in our present social structure, only great corporations have the resources and can develop the know-how to deal in a constructive manner with the environmental problem of technological societies."[16]

A month later, the faculty and students passed a plebiscite vote supporting Dubos and asked the trustees to vote the University proxy at GM for the "non-management slate of propositions demanding greater corporate attention to environmental problems." During one of the most dramatic days on Rockefeller's staid campus, a community strike and a teach-in protested the political and economic implications of the Vietnamese war, the tragic deaths of student protesters at Kent State University, and the role of the University in the march of events, specifically Campaign GM. A statement from the trustees was read, saying Campaign GM had become "so political in nature that it is advisable for the University in its corporate capacity to abstain from voting its shares of GM." The board decided the nation's environmental needs had to be based upon cooperative efforts among government, academic, and industrial groups, because "no one company or industry can accomplish this task." In deference to Dubos, but without naming him, the trustees said they would "offer the University's assistance, where relevant, to GM in trying to solve these problems, which must be solved aggressively and at an early date." Campaign GM failed ten days later during the General Motors annual meeting.[17]

These campaigns suggest a genuine naivety of Dubos about politics. In reality, there was something more behind this behavior. He knew himself well enough, while coping with a lifetime of disease, to refrain from engaging in controversy or arousing hostile feelings, even in the laboratory or scientific conferences. When confronted with strident ideas or people, he would fall silent and walk away. As stressful as these "visible scientist" activities were for him physically and emotionally, it seems clear he saw them in a different light. Just as he had perceived obstacles in creating a new science of human nature, he now learned firsthand why solving environmental problems would also need organizations, methods, and social forces that were currently outside established traditions and structures.

Dubos gave freely of his advice and the use of his name to environmental organizations, although remaining very selective in what he gave of himself, even to colleagues, or what he chose to do. "I act on what I believe to be worthwhile causes," he told John Adams, president of the Natural Resources Defense Council, adding, "I intend to remain a freelance of the environment movement even if it causes me some difficulty . . . [and] shall to the end try to avoid the tunnel vision which seems to affect many leaders of the environmental movement and may well, before long, alienate the movement from society."[18]

He contributed where he felt most confident and comfortable, by providing underlying attitudes and missions and helping to formulate new directions and policies. Not surprisingly, these activities were mostly in New York City, the place where he lived. To act on what he considered "one of the gravest threats facing humankind," namely, teenage unemployment, he joined the National Commission on Resources for Youth, founded by Judge Mary Kohler, which facilitated the active, creative participation of teenagers in their own communities.[19] In addition, as a founding member or trustee, he served as an unpaid consultant to several organizations.[20]

The stresses of extricating himself from the pressures and commitments brought on by the Prize were followed in 1971 by a near fatal attack of subacute bacterial endocarditis. This illness was the most feared complication of his rheumatic heart disease, one that was uniformly fatal before antibiotics. During a month-long hospitalization, while receiving massive intravenous antibiotic therapy, he seriously reconsidered the stresses that public life was causing.

At this point, coinciding with the closing of his laboratory, he resigned as editor of *The Journal of Experimental Medicine*, saying he was no longer interested in the science, and decided to "withdraw progressively from the world, as gracefully as I can."[21] He regretted not being vigorous enough to march with the students who were acting as spearheads of the environmental movement, since he deeply believed in grassroots efforts. What he did instead was appeal to their consciences not by actions but through words. His personal justification for these less visible efforts was that the environmental movement would soon need to go beyond scientific, technical, and political problems "to develop a philosophy of civilization."[22] These efforts, which consumed his remaining work days, form the substance of this chapter.

A Despairing Optimist Emerges

Around 1970, Dubos created a persona to temper his gloomy predictions with more hopeful visions. He introduced "the despairing optimist" in *Reason Awake* and went on to define this persona as someone who laments the deterioration of human values yet retains faith in man's potentialities. He used "despairing optimist" for a decade as the title of a regular column in *The American Scholar*; its theme—"where human beings are concerned, trend is not destiny"—has become one of his most famous mottoes.[23] Ecological thinking and problem solving were no longer advocated as exclusive research tasks for scientists and technologists. They were put forth as tasks for everyone to explore and apply to their individual lives.

As a despairing optimist, Dubos stood apart from, and even at odds with, other environmentalists. While they offered analyses of diseases afflicting earth and proffered vast technological and political controls to cure them, he attempted to identify what made the earth healthy and advocated human responsibility for finding paths to recovery and health. He disparaged environmentalists who called for impact and policy statements from "blue ribbon panels" of academic and scientific experts or for more legislation. His arguments for saving the earth called for integrating all human activities, not just fixing the ruins. There was a flair and determination in his appeal that "If we really want to do something for the health of our planet, the best place to start is in the streets, fields, roads, rivers, marshes and coastlines of our communities." How can we preserve our own health or think of saving the earth, he lectured, if we cannot do something about the place where we live?[24]

An early success was close to home in New York City's Jamaica Bay. Urban development had transformed this wetland into a wasteland: once a home for oyster beds, a spawning ground for fish, and a waterfowl sanctuary, it had become a nearly lifeless body of water whose shoreline was almost entirely landfill. More than fifteen hundred outflow pipes discharged raw sewage into the water. While area residents began fighting for a cleanup of Jamaica Bay, a lone Parks Department horticulturist, Herbert Johnson, was working to find grasses, shrubs, and trees that would colonize the landfill.[25]

A turning point for Jamaica Bay came in 1970 when Dubos and several dedicated local citizens mobilized public opinion to block extension of runways at John F. Kennedy International Airport into the bay. His contribution, hailed in *The New York Times*, was to ridicule the idea that a two-year study was needed to determine whether this expansion would damage the bay's ecosystem. Common sense, he declared, left no doubt as to the adverse impact of sewage, dredging, and increasing landfill for more runways.[26]

After the airport expansion was defeated, more improvements came from sewage treatment plants to clean the water and from converting nearly half the bay's perimeter to parkland. As a result, Jamaica Bay Wildlife Refuge, within Gateway National Recreation Area, has been repopulated by three hundred species of birds as well as mollusks and fish. In 1987, the City of New York named a thirty-six-acre saltwater marsh projecting from the southern shore of the bay, once an automobile junkyard, the Dubos Point Wildlife Sanctuary.

Jamaica Bay's success exemplified many facets of what Dubos was promoting. Its problems did not need the expertise of a scientist, analysis of an economist, or wiliness of a politician to know that what happened was due to careless and unsound biological behavior. Restoring the bay was a social issue dependent on value judgments and actions by motivated individuals. Since wetlands are biologically resilient, once the sources of pollution were removed, the natural landscape recovered and native flora and fauna returned. As Dubos once exulted over this victory, "If I were Billy Graham, I would go out and preach to people that the best way to save their souls is to save the environment of cities like New York."[27]

Another one of his ventures into saving a New York environment has become a posthumous success. On his first evening in the United States in 1924, Dubos tried to take a walk along the Raritan River in New Jersey. To his great surprise and disappointment on this and many later occasions, American river banks, unlike many in Europe, were inaccessible to pedestrians. When he became a Manhattan dweller, he hoped in vain that Herman Melville's portrayal in *Moby Dick* was still true and that "right and left, all the streets take you waterward." Finally, after accepting for too long that New York City's nearly six hundred miles of waterfront were grossly mismanaged, spoiled, and mostly inaccessible, he spoke out in a 1980 op-ed essay for *The New York Times*, lamenting the unexploited potential of American waterfronts and pleading for waterscapes that would be easily accessible on foot, scenically attractive, and engage people in various outdoor pleasures. This commentary led to his appointment to the Citizens for Balanced Transportation that fostered Westway, a park and parkway along the Hudson River. As the only ecologist who favored a combined multibillion dollar highway-park-commercial development, he chided the city for failing to take advantage of the scenic diversity of its waterfronts.[28] Nonetheless, environmentalists raised so many concerns about endangering fish species and taking away funds from public transportation that the project was abandoned. Dubos admitted he would "not live long enough to enjoy the Westway park. But I have enough social conscience and pride to believe that adding to the future beauty of the New York City waterfronts is more important than making my subway ride more comfortable. . . . Opposing Westway is the equivalent of closing libraries to save money—the evidence of a loss of intellectual values in the American mind."[29] Now, two decades after Westway was stopped, a grand yet modified revival designated Hudson River Park is transforming five miles of Manhattan's western shoreline. When completed, it will be the largest riverfront park in the United States, leading the governor and mayor to declare in 1996 that this waterfront project was "a demonstration that people should not give up."[30]

All the environmental messages of this despairing optimist were taken from his most basic *biological* lesson, repeatedly demonstrated, that living things are inextricably linked within their environment. In brief, this lesson teaches that humans and landscapes, like soil microbes, are characterized by resilience, flexibility, and diversity, so that each one has the potential to express many traits and to function in diverse capacities. What an individual or landscape becomes depends on the conditions under which it develops. Failed adaptations, whether by humans, landscapes, or microbes, result in disease, waste, or devastation; successful adaptations can foster and restore health.

A New Environmental Attitude

Dubos spent the final ten years of his life trying to relate human health to that of the whole earth. To call his idea of collective ecological well-being a theory or philosophy is too grand, especially for someone who consciously avoided giving names or labels to anything that might be construed as a big notion, a sweeping doctrine, or a philosophical system. "I never speak as a philosopher," he said, "only as an experimental scientist who introduces humanistic considerations into all scientific and technical problems."[31] His messages favored practicality and function over policy or ideology, emphasized health as an ongoing activity, and evaluated success by the extent an individual or society meets the unpredictable circumstances of life. His goal was not to provide answers but to make individuals aware of how they could shape their own health and destiny and how this in turn would affect the health of the earth.

The first phase involved defining two positions from which he viewed ecological well-being: the "real" environmental problem and the word *environment.*

The real predicament, or survival is not enough. The very first Earth Day, 22 April 1970, was a day unlike any other. Despite the idea of "celebrating" the Earth, the aims were solemn. Its originator was Gaylord Nelson, senator from Wisconsin, who suggested a day be devoted to an environmental teach-in.[32] Nearly twenty million Americans took part in the day's teach-ins, bike-ins, dump-ins, sit-ins, and ecology fairs. Most orators that day warned that the year 2000 would not be the dawn of a technological utopia but instead a gloomy sunset for many forms of life, especially human life. In their midst, Dubos offered a lone view that the "real" environmental problem was not destruction of life but its progressive degradation and disorder, and not death but rather a worthless existence for human beings.

Two months before Earth Day, a *New York Times* story quoted Dubos' early warning to the New York City planners that the day would be a "glamorous event and stupendous failure" unless it was backed by a long-range protection program.[33] Furthermore, he predicted the single-issue environmental projects would be forgotten "when you become vice-president of the local bank, as many of you will." Dubos was classified as an "eco-philosopher" alongside Henry Thoreau and Henry Beston in *The Environmental Handbook*, a source book prepared by Friends of the Earth that went through three printings in advance of Earth Day.[34] During Earth Week, Senator Nelson, Margaret Mead, Paul Ehrlich, Ian McHarg, Barry Commoner, and Dubos were the environmental advocates who were interviewed by Hugh Downs on national television's *Today Show*. Their impact was intensified when these interviews were published immediately as *New World or No World.*[35]

On Earth Day itself, Dubos diagnosed wasted resources, pollution, and ruined landscapes as symptoms of a much larger and invidious disease—the dehumanization of mankind. Speaking on two college campuses that day, he told students that "modern societies can survive only if they outgrow the growth myth and if they make quality of life, rather than quantity of production, the criteria of their success." Citing Campaign GM, he noted that General Motors had a social duty to forego more and bigger automobiles and to build "a system of transportation better suited to the needs of American life."[36]

A few months later, his role as a celebrity environmentalist was secured with editorials in two mass circulation magazines, *Life* and *Reader's Digest.* "Mere Survival is Not Enough for Man" made clear that while other environmentalists were fighting *for* preserving environments, he believed that we "may be doomed to survive as something less than human . . . in the polluted cage of technological civilization." The editorial optimistically concluded that individuals could design new environments to help express "attributes that make human life different from animal life."[37]

Among many other lectures, Dubos told scientists and technologists in London that "The Predicament of Man" originated from their gross neglect of human attributes. Opposing social scientists Jacques Ellul in France and John K. Galbraith in the United States, who were arguing technology was enslaving mankind, Dubos said the destructive demon of scientific technology is "in those men . . . who are more interested in things than in conditions suitable for the development of human potentialities." By simply correcting environmental defects with technological fixes, he warned "we behave like hunted creatures" escaping dangers by taking shelter behind endless protective devices—"today afterburners on our cars and complicated sewage treatment systems, tomorrow gas masks and filters on our water faucets." Reminiscent of warnings about antibiotics, he predicted technological fixes would not get to the source of problems, have only temporary usefulness, and "increasingly complicate our life and ruin its quality."[38]

Whether audiences were teachers, architects, museum curators, or psychiatrists, he evoked experiences of daily life to sensitize them to the larger idea that a cumulative, sinister loss of their sensual perceptions was atrophying the quality of their life. We adapt so unconsciously to our surroundings, he worried, that we no longer seem to mind the stench of automobile exhausts, addictive electronic media, ugly urban sprawl, starless skies, treeless avenues, shapeless buildings, tasteless bread, joyless celebrations. In comfortable built environments, surrounded by plants and animals we choose, without knowing either our sources of food and water or the destinations of our garbage and sewage, we feel we have escaped nature.[39]

As he saw it, the "real" environmental problem is that we have become biologically adrift and are practicing "biological warfare against nature, ourselves, and especially our descendants." Instead of treating symptoms and ill effects produced by these objectionable conditions, he advocated two solutions: retaining as consciously and fully as possible the direct perception of reality and changing the conditions themselves before the crises occurred.

The real meaning of environment, or sustainability is not enough. To amplify why the "real" predicament was not better understood, Dubos produced a far-reaching definition of *environment* that conveys how the *total* environment—the obvious physical forces, the less obvious social forces created by the community, the almost neglected perceptual forces intercepted by sense organs, and conceptual forces created and existing only in individual minds—participates in human well-being.[40] "If you want to study life," he once said, "study the environment."[41]

By training, Dubos had been shaped by three nineteenth-century biological theories: Pasteur's germ theory of disease, Darwin's theory of evolution, and Claude Bernard's theory of homeostasis. Just as he earlier supplanted the germ theory with more ecological views of disease, showing a microbe is necessary but not sufficient to cause disease, he now challenged the theories of evolution and homeostasis as inadequate for understanding the environment as a dynamic force.

As a biologist, he accepted scientific evidence of evolution whereby mechanisms of natural selection and mutation gradually incorporate genetic changes in human beings. As an ecologist, however, he now argued these processes contributed only to evolution of *species* but offered little to account for mechanisms by which *individual organisms*—humans—go beyond perpetuating and satisfying themselves and begin to imagine, and create, their social and cultural environments. A frequent statement in many essays was that genes do not determine traits, only govern responses of the organism to the environment. Evolutionary biologist Stephen Jay Gould also argued that Darwin himself "tiptoed lightly" when it came to extending his theory of evolution to *Homo sapiens*. "Although Darwinism surely explains many universal features of human form and behavior," Gould added, "we cannot invoke natural selection as the controlling cause of our cultural changes."[42]

By coincidence, at the time Dubos was redefining the word *environment*, geneticist and Nobel laureate Salvador Luria asked him to critique the manuscript of his autobiography, *Life: The Unfinished Experiment*.[43] The detailed response to Luria divulged his belief that while evolutionary theory may explain continuity from one life form to another, it could not account for "the evolutionary passage from matter to life and from life to consciousness." Aware of his uncertainties, Dubos expressed what, for him, was a black idea: "Theoretical biology is now in the same intellectual position as Newtonian physics was at the end of the last century—highly effective in describing the mechanics of events, but essentially naive in discussing their causation and nature."[44] Moreover, he said he refused to express or defend these doubts publicly, because such "an opinion from a biologist might be taken as an expression—not of doubt but of belief in some hoary vitalism," which he believed would discredit science. What he espoused elsewhere and openly, however, was the conviction that the uniqueness of "man is characterized not by his biological endowments but rather by what he has created from them," adding that "adaptation is not the exclusive property of modern geneticists."[45] In other words, to recast his own saying, where human beings are concerned, biology is not destiny.

The theory of homeostasis was also criticized by Dubos as too simplistic for understanding the environment. As a biologist, he once counted himself as accepting this feedback system by which an organism resists distorting environmental stimuli, makes self-corrections and repairs, and returns the body to a constant state of health. As an ecologist, however, he began to find that Claude Bernard's fixity of the *milieu intérieur*, Lawrence J. Henderson's "fitness of the environment" (1913), Walter Cannon's "wisdom of the body" that Cannon termed "homeostasis" (1932), and Norbert Wiener's self-correcting cybernetics (1948) could not account for frequent failure, illness, and disease.[46] He relegated homeostasis to mechanisms of biological security and awareness that emerged during our evolutionary development, citing the example that an immune response protected "primitive man against wounds and infections, but it has not caught up with the modern surgeon's boldness."[47] He argued instead that there is no fixed equilibrium, no status quo, and what was being overlooked is how the very process of living transforms environment and organism irreversibly.

In passing, Dubos would have almost certainly objected to the current environmental watchword *sustainability* as another form of homeostasis. Sustainability was initially conceived as a political and economic policy for the United Nations in 1987. Like the vague uses of *ecology*, the term has now been applied to all kinds of environmental issues with emphasis on limits to growth, conservation, and an equilibrium or steady state. Dubos believed that "if the ecologists' concept of man's relation to the total environment really did imply a steady-state system, ecological philosophy would be indeed dangerous as well as wrong," because individual development and true evolution would be impossible.[48] Stephen Boyden, an early member of Dubos' tuberculosis laboratory and founder of the Nature and Society Forum, observed recently that "sustainability is the bottom line," for without it a society could not long survive. More importantly, Boyden observed, "We should surely be aiming, and hoping for, something much better than mere survival—mere sustainability" and he called for a society that continuously develops its rich, diverse, and productive natural environment. Our goal, he added, "should be ecological well-being."[49]

The concept of environment was newly defined by Dubos as *homeokinesis*.[50] This unusual term, from the Greek noun *kinesis* for movement, was one he likely coined himself to oppose the word homeostasis. By reasoning that the theories of evolution and homeostasis implied instinctive *reactions* that are passive, blind, and often faulty, he invested in *homeokinesis* the idea of making conscious *responses* that involve awareness, motivation, and free will. *Homeokinesis* conveys a dynamic relationship between the external and internal environments that embodies all degrees of exchange, flux, and integration. This interpretation of total environment was his effort to supersede evolution's purely biological struggle for survival and homeostasis' instinct for sustainability with a supremely human willed effort to adapt consciously—even creatively.

Concluding Visions of a Wise Ecologist

These two definitions, which put forth a new attitude toward the environment, set the stage for Dubos to implement the notion of *homeokinesis*. He did this by devising the kinds of "patterns of response," or unexplored "biological truths," that he earlier advised would advance a humanistic biology. Each concept bears an indelible stamp of his originality: think globally, act locally; improving on nature; and making creative adaptations. Their notions of integration, accommodation, and creation still counter notions of antagonism, chaos, and destruction that are exploited by other ecologists and environmentalists. Although they represent his final thoughts, they are in no sense exhaustive or finished. For clarity, these concepts are presented separately. In reality, they were never separated in his mind, and their inquiry overlapped as elements of a whole vision to facilitate salutary relationships between nature, technology, and human beings—ecological well-being.

Think globally, act locally. Without a doubt, the most famous motto of the twentieth-century environmental movement was "Think globally, act locally." This now-iconic phrase has galvanized innumerable crusades by grassroots activists, nature organizations, and political campaigns and has proliferated on bumper stickers, billboards, lapel buttons, and tee-shirts. It spread so rapidly and universally that very few people know it originated with Dubos.

The idea for this phrase emerged as the world moved from the local euphoria generated by the first Earth Day to global environmental challenges presented at the United Nations Conference on the Human Environment in 1972. British economist Barbara Ward and Dubos provided the conceptual framework for this international gathering by preparing the unofficial, privately commissioned report known as *Only One Earth: The Care and Maintenance of a Small Planet*.

Maurice Strong, Secretary-General of the Conference, was attracted by Dubos' human approach to environmental problems and asked him to join a group of international experts who had already been working under Ward.[51]

A draft was nearly ready, but neither Ward nor Strong was satisfied. Dubos responded with detachment since he was still in bed weakened by the bout of endocarditis mentioned earlier. By late September he felt well enough to tell Ward that her *tour de force* was too focused on problems in terms of emergencies and needed more optimistic views to emphasize how humans must take responsibility for nature and how, even though nature is resilient, humans can reverse trends. This attitude became the basis of a completely revised report that doubled in size.[52]

After the new draft was reviewed by 152 consultants from fifty-eight countries, Ward and Dubos took up the sensitive task of incorporating the disparate native values of these individuals and nations into a global perspective about the status of the human environment. In keeping with the report's mission, the authors did not propose specific actions, suggest resolutions, or offer utopian solutions. Their purpose was to provide background information relevant to official policy decisions. They placed humans in "the hinge of history" and concluded that a planetary strategy for their survival must be based on their cumulative efforts in countless small political and social activities. This theme planted seeds of the famous motto.[53]

Ward and Dubos, to the surprise of many, did not participate in the official U.N. meeting. They joined a small forum in Stockholm, also organized by Strong, where they spoke freely and without national interests on environmental topics of personal interest. "Unity through Diversity" was Dubos' topic, in which he dared articulate his opinion that the real conflicts at global conferences are not about scientific "knowledge or interpretation of facts but . . . differences in the value judgments they put on these facts." Moreover, he predicted these conflicts were inevitable because all humans have not only the same biological needs that act as a unifying (or global) force of the world but also an immense variety of social and individual demands (or local forces) that often lead to discord. One environment could therefore never be equally desirable for all individuals even if basic needs were satisfied. A summary statement from this lecture lies at the origin of his famous slogan: "In practice, a global approach is needed when dealing with the problems of the spaceship earth which affect all of mankind. But local solutions, inevitably conditioned by local interests, are required for the problems peculiar to each human settlement."[54]

Before the motto was crystallized into four words in 1978, Dubos strengthened his views about human attachments to regions and places (discussed in the next section). Consequently, he became more aware that thinking globally was not natural to human beings and that international agreements are so "generalized beyond any operational usefulness" that they could not substitute for local action.

Several versions were composed to pair "think globally" with "act locally." The first known appearance of the slogan joins the two halves with the conjunction *but*, which he always preferred for its stress on personal involvement. In 1975, commenting on the futility of solving environmental problems by legislation, he advised Inform, an organization studying corporate responsibility, that "we may try to think globally or on a national scale, but we, if we want to be really effective, have to act locally. . . . Legislation is almost meaningless if it is not implemented through local action; and local action is possible only to the extent that it reflects grassroots attitudes."[55] Also, in a letter to Paul Ehrlich two years later, he wrote, "I do not share your optimism about global control of any problem. In my opinion, the best we can do is think globally but we can only act locally."[56]

Other versions connected the phrase with *and* or *versus*. In a 1977 "Despairing Optimist" column, he wrote "the ideal is to think globally and to act locally," suggesting both activities are potentially equal, although the article emphasized local action.[57] A more controversial, pessimistic version, "thinking globally versus functioning locally," was an effort to counteract some undesirable trends of globalization toward standardization and monotony with a new frontier spirit in America to identify with the place where we live, not cerebrally but through direct experiences to satisfy biological senses of belonging.[58]

For the record, the title of an interview in 1978 with Dubos in the *EPA Journal* is the first known appearance of the four-word slogan, "Think globally, act locally."[59] Criticizing students who discussed saving the globe while ignoring the messiness of their cafeterias, he said, "It is a good intellectual exercise, but the only way where you can do something is in your own locality. So think globally, but act locally." A year later, the slogan became the headline of a four-page advertisement by SmithKline Corporation in the *Wall Street Journal* and *Newsweek*, featuring color photographs of Dubos, his personal commentary, and his prediction that "There is a better chance for creativity, safety, and manageability in multiple, fairly small systems, aware and tolerant of each other, but jealous of their autonomy."[60]

"Think globally, act locally" was used cautiously by Dubos. He had a general mistrust of the facile nature of slogans and hesitated to boast about creating this one.[61] There were, however, two places where he enhanced its ethical implications. In *The Wooing of Earth* (1980), the phrase marks his crowning view of a human ecology in which "global thinking and local action both require understanding of ecological systems." Ecology, he said, "is nothing more than the study of interrelationships between living things and their environment; it is therefore ethically neutral. These relationships, however, are always influenced by the human presence, which introduces an ethical component into all environmental problems . . . ecological thinking must be supplemented by humanistic value judgments."[62] A chapter of *Celebrations of Life* (1981) called "Think Globally, But Act Locally" concluded the earth is merely "a physical system for the support of life, whereas the words 'place' or 'nation' denote an environment which has been emotionally transformed by feelings . . . [and] enters into the substance of [our] lives."[63]

The legacy of "Think globally, act locally" resides in its spirit of practicality infused with a sense of environmental citizenship. Between abstract awareness and concrete action, between imagination and experience, lay its decree that earth housekeeping and earth health begin at home. If an ecological conscience begins at home, then there is hope that every individual can make a difference.

Improving on nature, or humanizing the earth. In late 1972, Dubos began to espouse the view, shocking to twentieth-century environmentalists, that humans can improve on nature. Only a century earlier, this view had been widespread among American conservationists. A popular example can be found in the essays of Hudson River naturalist John Burroughs, who was admired by Dubos for his willingness to disturb nature. During the Civil War, George Perkins Marsh, scholar, public servant from Vermont, and pioneer of the modern conservation movement, reviled destruction, waste, menace, and possible extinction of human life in his book *Man and Nature, or Physical Geography as Modified by Human Action* (1864). Marsh was also optimistic that "the cost of one year's warfare" could procure for "almost every country that man has exhausted, an amelioration of climate, a renovated fertility of soil, and a general physical improvement, which might almost be characterized as a new creation."[64]

Dubos actually borrowed the phrase from conservationist President Theodore Roosevelt, who told a gathering of governors in 1908 that "man can improve on nature by compelling the resources to renew and even reconstruct themselves in such a manner as to serve increasingly beneficial uses."[65] Quick to recognize how contentious this phrase sounded to twentieth-century environmentalists, Dubos always added that improving depends on ecological wisdom and social will.[66]

At the same time Dubos was working on *Only One Earth*, with its global perspective on human environments, he was writing *A God Within* (1972). The book is not a theological text, as some infer from the title, but an exploration of cultivating a distinctive spirit or genius of each place, social group, and person. Its title and theme come from Pasteur who called enthusiasm, from the Greek *en theos* or the god within, the "most beautiful word" because "the grandeur of human actions is measured by the inspirations from which they spring . . . the living sources of great thoughts and great acts."[67]

What was proposed in *So Human an Animal* as working with nature, and extended in *A God Within* as revealing hidden qualities in nature, was soon thereafter transformed again into directly managed care of nature. This happened immediately after the Stockholm meeting when the Duboses traveled around Sweden, Greece, and his childhood villages in the Île de France. In each region he became aware of how humans had shaped these landscapes for so long that they were no longer original or primitive, yet they signified the "nature" celebrated by their citizens, artists, and poets. During a visit to the country home of Swedish Prime Minister Olof Palme, he learned that public funds were subsidizing farmers to keep fields open; the Swedes thought nature was being ruined by brush invading abandoned fields. "That's not Swedish nature," Dubos told his startled hosts, "Swedish nature was just trees . . . all that we call nature is really something created by human enterprise." What the Swedes were preserving was "man-made" countryside![68]

Within six months of his return to New York, he began studying environmental success stories and writing about "Humanizing the Earth." He did not dismiss environmental disasters or vanquishing wilderness as he introduced a new level of creating environments that are "ecologically stable, economically profitable, esthetically rewarding, and favorable to the continued growth of civilization."[69]

He readily accepted lecture invitations from such varied audiences as the state legislature of Vermont, Greek city planners, and French botanists so that he could explore with each group some specific ideas for their local solutions to improve on nature. Some lectures resonated so strongly that people still mistakenly think Dubos was a trained architect for relating conscious design to daily habits, landscapes, buildings, and cities. He advocated spaces that would satisfy such biological needs as open horizons, protective enclosures, a village atmosphere, and natural surroundings. He specified places or "stages" that could intensify human encounters and allow citizens to become participants rather than observers in community events and ceremonies. Other ideas were suggested for improving agriculture patterns, creating safer industrial and medicinal chemicals, finding more suitable ground cover to replace lawns, and planting more interesting trees to relieve urban monotony. Audiences were frequently shocked to hear Dubos pronounce lawns "unecological." As a Manhattanite who loved to watch bicycle riders in Central Park, he promoted the "necessary pleasures of park life," expanding on Frederick Law Olmstead's view that even the illusion of nature in the city allows inspirations and escapes from the urban "cage of reality." Although examples in these lectures changed, the message was the same: cultivate environments that would nourish such human biological needs as tranquility, beauty, involvement, fertility, and inspiration.[70]

In 1976, Dubos won the Tyler Prize, a new prize at the time but now regarded as a world class prize in environment and ecology. The award recognized his five decades of concern with the effects of environmental forces, his influential ecological philosophy, and his current campaign to "counter balance . . . those who think that man is rapidly ruining the earth." The *Los Angeles Times* captured the shocking effect of improving on nature in a front-page profile of Dubos as he arrived to accept the award. Dubos playfully exaggerated the reporter's potentially hostile attitude, commenting "What I'm going to say is so wild I have to leave town at 4 o'clock." The news story referred to him as a "respected maverick," "75-year old rabble rouser," "renegade," and "traitor among environmentalists." Nevertheless, the reporter described the impact of his lecture as producing "more awe than criticism."[71]

As he had anticipated, environmentalists criticized improving on nature, specifically those who regarded humans as intruders and destroyers of earth and those who defended preserving wilderness.

In 1978, population biologist Paul Ehrlich and social historian Leo Marx worried that Dubos might give people the impression that environmental destruction was less important than they had been led to believe and that degraded ecosystems would easily regenerate. Ehrlich was "nervous that some mineral exploration company will soon have ads claiming that René Dubos says it's all right to clean the planet right down to bedrock because any time we stop the natural ecosystems will be restored to their full glory, in a few years!"[72] Marx found Dubos "complacent and conservative" for not giving adequate attention to the explosive character of urban-industrial economies and acquisitive capitalism. As a result, Marx said, Dubos "serves those who feel that everything will be all right if we simply go along, picking up a few beer cans, and denying the need for any systemic change."[73]

In a further exchange, Dubos told Marx their difference of opinion stemmed from the fact that he was emphasizing, as scientist, the "biological mechanisms of ecological recovery and not its social determinants" and that he had "some faith in the possibility of changing public attitudes and thereby changing public policies." To which Marx replied, the idea of changing public attitudes "is a laudable aim, although I am skeptical."[74]

Wilderness advocates faulted the idea of improving on nature as "beneficent pastoralism." Environmental historian Roderick Nash criticized Dubos and poet-farmer Wendell Berry for being "proponents of the garden-earth scenarios," similar to what Marx called "the middle landscape." Noting Dubos' French origins and pastoral biases, Nash commented, "With regard to land use, Dubos does not believe in wild nature as a criterion. . . . He thinks of a gardenlike earth, shaped and controlled by man, as the logical fulfillment of the human potential." Nash argued that wilderness is just as dead in the garden scenario as in the concrete wasteland, although ending beneficently as Dubos proposed rather than destructively. But this "middle, rural option," he concluded, "may in fact be the worst of both worlds, lacking both elk and computers."[75]

In 1979, *Geo Magazine* asked Dubos, a defender of humanizing the earth, and Edward Abbey, the arch defender of wilderness where no one enters, to present their arguments side by side. Dubos distilled his earlier message from the Tyler Prize lecture that "the earth is to be seen neither as an ecosystem to be preserved unchanged nor as a quarry to be exploited for selfish and short-range economic reasons . . . humankind and the earth [are] a symbiotic system undergoing adaptive changes that result in a continuous process of evolutionary creation."[76] Abbey, while empathetic to the attractive, fruitful rural scenes praised by Dubos, challenged him to reconsider what had been lost in this process and stressed that wilderness is not a luxury but the essence of personal liberty, independence, and dignity, so "if we allow the freedom of the wilderness to be taken from us the very idea of freedom may die with it."[77]

There were also defenders of "improving on nature." Conservation Foundation president William K. Reilly knew Dubos was "outside the mainline of American conservationist opinion." Nonetheless, he invited him to join the foundation's board. Dubos demurred and told Reilly, "My opinions are not accepted by most American environmentalists, you know." Reilly prevailed, observing that "his uniquely valuable and important" voice was needed "to deepen our understanding not only of nature but of ourselves and of our own power and ability to alter our environmental destiny and give it a benign and beautiful shape." Reilly believed American conservation had provided a charter for wilderness preservation but had "left us timid and confused in the exurbs and suburbs. Our tradition has taught us confidently to preserve but left us uncertain about our capacity to create. Dubos stood for creation."[78]

An especially thoughtful response by Dubos to those accusing him of having betrayed the cause of the environment came in a 1978 lecture celebrating the thirtieth-anniversary of the publication of Aldo Leopold's *A Sand County Almanac*.[79] John E. Ross, professor of environmental studies at the

University of Wisconsin, who organized the event, said the lecture hall was overflowing with several hundred standing in the aisles. Even though his "presentation was counter to much of the environmental thinking in Madison at the time," and there were some hostile questions, twenty years after this lecture he observed that "his philosophy has considerable staying power."[80]

Just as Dubos had been profoundly influenced by Lewis Mumford who "seizes any occasion to preach the rehumanization of life,"[81] he now found himself a kindred spirit of conservationist-ecologist Aldo Leopold. In his lecture he presented Leopold as one who advocated rehumanizing the earth, despite his role as a founding member of the Wilderness Society. His opening comment challenged the skeptical audience to think differently: "Why do you admire that ecological disaster and want to visit that celebrated example of soil erosion, the Grand Canyon?" He went on to suggest leaving behind nostalgic views of nature and to rethink Leopold's views of wilderness, land use, and land health.

Like Leopold and Abbey, Dubos defended wilderness as "essential to our survival and sanity," but, he argued "the concept of wilderness has itself been humanized and is now as much a state of mind as an expression of Nature." More to the point, he added, "Leopold does not state anywhere in his book that the natural biotic communities of wilderness are necessarily the most desirable and that it is wrong in principle to alter them." Therefore, he suggested it was time to redefine nature, since "we cannot help but be involved in the management of nature," including wilderness.[82]

Recognizing Leopold's land ethic as the "great commandment of conservation philosophy," Dubos agreed that each question of land use must be judged ethically, esthetically, and economically, thus echoing Leopold's famous words, "A thing is right when it tends to preserve the integrity, stability, and beauty of the biotic community."[83] Nevertheless, Dubos deemed this land ethic was misinterpreted when people took a static view toward the environment and hesitated to change anything for fear of ruining integrity, stability, and beauty. "The land ethic does not rule out the transformation of nature," and he pointed out that Leopold planted "lots of trees" to heal his abandoned farm. "He was changing the landscape, changing it in ways that he considered proper. But this is of course what we all must do." Defending his own efforts in the Hudson Highlands, he said, "the landscape designer, without ruining the ecology of the region, can cut a few trees, open space along those wonderful rock formations, then all of a sudden . . . find a kind of architectural beauty inherent in the land that could not be seen before." Leopold's land ethic, rooted as it is in love of the land, was in effect described by Dubos as another form of improving on nature for showing how humans can generate new environments, new phenomena, and new values that are not predictable from a deterministic order of nature.

In many ways, conservation for both Leopold and Dubos implied human intervention with an obligation to reverse abuse and to harness nature's powers of recovery. This attitude respects Marsh's nineteenth-century perspective on humanity's role as "co-worker with Nature in the restoration of disturbed harmonies."[84] By stressing how Leopold "pleads only for concern with ecological health," Dubos identified with his philosophy of conservation to cultivate the healthy functioning of the biotic mechanism, whether it pertains to creatures, landscapes, or communities.

On the tenth anniversary of Earth Day, in April 1980, Dubos reinforced this philosophy of improving on nature in a lecture at Princeton University. Although somewhat superficial, he sketched "Five E's of Environmental Management" and offered concrete and deliberately startling ideas for human efforts. These categories included ecology (remove lawns); economics (appreciate the value of seeing the Milky Way in New York City); energetics (abolish free-standing houses); esthetics (create more greenbelts and parklands in the suburbs); and ethics (keep landscapes healthy by preserving their capacity of self-recovery).[85]

Many of these unorthodox ideas were enlarged in Dubos' penultimate book, *The Wooing of Earth*.[86] Getting them into book form was more difficult than he first imagined and took six years, longer than any other book. He contemplated an illustrated book of landscapes "that have become better by being nicely treated by man,"[87] rejected a collaboration with the "flying photographer" Georg Gerster, and finally wrote a historical essay of environmental success stories based on his experiences and observations.[88] Despite editors who wanted a scholarly argument, he refused, saying its mood and organization "does not argue but rather it almost states visions" of humanizing the earth.[89] He at first preferred a forceful French phrase *la terre a besoin des hommes* (earth needs humans) to convey the human role in cultivating land health, but, as he had learned from finding inherent beauty on his own farmland, it takes more than work; it takes love. So he borrowed the gentle words of poet and Nobel laureate Rabindranath Tagore, who in 1926 portrayed this human task as an "active wooing of the earth."[90] In the end, he championed humans whose privileged biological place in nature entails an obligation to practice stewardship to improve on nature.

When the French translation of this book won the Prix Sully–Olivier de Serres, President Valéry Giscard d'Estaing saluted Dubos in person for espousing an "écologie civilisatrice," remarking that ecology was no longer only a defensive effort but now a civilizing force "where nature needs man who can alone reestablish the natural equilibrium compromised sometimes by his own actions."[91]

Making creative adaptations. The two biotic concepts—Dubos' human health and Leopold's land health—are natural counterparts, and they are brilliantly captured in Leopold's analogy that "in land, just as in the human body, the symptoms may lie in one organ and the cause in another. The practices we now call conservation are, to a large extent, local alleviations of biotic pain. They are necessary, but they must not be confused with cures. The art of land doctoring is being practiced with vigor, but the science of land health is yet to be born."[92] Leopold once defined *conservation*, or land healing, as "a positive exercise of skill and insight, not merely a negative exercise of abstinence or caution."[93] Dubos described the human counterpart of conservation or healing as "creative adaptations."

While "improving on nature" focused on how humans could influence external environments, creative adaptations was also concerned with "improving on self" in relation to nature. This idea is ostensibly Dubos' most trenchant and optimistic contribution to ecological well-being. Trenchant, because it encourages change, proscribes passivity, and bypasses symptoms to get at the source of problems. Optimistic, because it sanctions hidden potentialities of human energy and imagination, expects newly created resources, and prepares for new attitudes and values.

"Creative adaptations" addressed concerns of those who were beginning to despair that science cannot solve all technological problems, that power and wealth do not necessarily make for a happy life, and that medicine cannot cure all disease. Some spoke of the times as The Age of Anxiety and others of The New Pessimism. In response, Dubos tried to allay these feelings by articulating aspects of creative adaptations that, if exercised, could restore personal control over deficiencies or life-threatening situations, and at the same time achieve and preserve health. The aspects described below include creative associations or symbiosis, deliberate choices, pursuit of happiness, self-healing, and creative evolution.

CREATIVE ASSOCIATIONS. *Creative associations*, or *symbiosis*, was Dubos' way of showing how relationships among individuals or between humans and nature could achieve something neither could alone. As he often did, he made abstractions concrete and accessible by using parables. In this case, lichens provided a favorite example of how a partnership between two organisms transcends biological utility and "adds something unexpected to the interest and richness of life."[94] Beginning with his Pasteur biography in 1950, a lichen parable appeared in nearly every book, in countless lectures, and even in a Sunday church sermon,[95] although it was not until 1962 that he discovered to his delight that the word *symbiosis*, or life together, had been coined in 1879 to explain mutual relationships of algae and fungi in lichens.[96]

He also learned from lichens and, in Thoreau's words, was a lichenist who "extracts nutriment from the very crust of the earth" and whose study is "evidence of titanic health, a sane earthiness."[97] Dubos spent hours gazing on these visible microbial bouquets growing on trees, boulders, and old stone fences in his Hudson Highlands hemlock forest, and from them he extracted many analogies to human behavior. As a result of living together, lichens produce thousands of beautiful, complex forms, colors, chemicals, and beneficial properties that are more than the sum of their parts. He saw they flourished in depleted Garrison landscapes yet perished in polluted landscapes of New York City. Over the years, he observed how their exact nature, like that of soil microbes, depends on environmental factors operating at any one time and how their nongenetic mechanisms of adaptation truly revealed what might be achieved with cooperation rather than struggle. Most remarkably, their mutual relationships seemed to go beyond biological survival or needs of either organism. Humans, he concluded, can also "become socially integrated and thus generate new social organisms whose characteristics transcend their own." Individuals, communities, regions, or nations can "shape the raw stuff of nature into patterns which integrate the materials provided by the wilderness with those of our human nature—a truly creative symbiotic process."[98]

DELIBERATE CHOICES. Dubos had lived in America for half a century and, except for letters to his family and a few lectures, he seldom wrote in French. Suddenly, or so it seemed in 1973, Albert Blanchard, a French publisher, asked him to write a book on human evolution in French. This was not an impulsive request. In 1970 French biologist and Nobel laureate Jacques Monod had produced a sensation with his book *Le hasard et la nécessité* (*Chance and Necessity*) with the message that evolution depends entirely on random mutations (chance) screened by natural selection (necessity) leading to an absurd, meaningless universe. Dubos knew this book and expressed his opposition to its arguments among colleagues but never in print. What he did in accepting Blanchard's offer was to present a completely different biological message, bypassing Monod, on human evolution. Blanchard's response to Dubos' synopsis was that his message would have a "privileged welcome in France" because "your approach and your conclusions will set you apart from his [Monod's] pessimism."[99]

Choisir d'être humain (*Choices that make us human*, published in English as *Beast or Angel*) was in many ways a calculated risk for Dubos, who reverted to a characteristic French spirit of reasoning and explored a highly debatable topic. With a certain finesse, neither philosophical nor scientific, Dubos turned away from debate and provided a historical essay on how "human life is the expression of deliberate choices based on the anticipation of the future . . . [making this] the root of creative adaptation."[100] *Chance*, in Dubos' thinking, "provides the conditions and materials" from which civilizations renew themselves, but it is a *necessity* of the human mind that "selects among these options and that organizes the raw stuff of nature."[101]

On this important question, Dubos differed sharply from evolutionary biologists who were dominated by interest in conflict and competition. "We are human to the extent that we are able and willing to make choices that enable us to transcend genetic and environmental determinism, and thus to participate in the continuous process of self creation." As he wrote to sociobiologist E.O. Wilson, after refusing to review his book *On Human Nature*, many attributes of the human species cannot be accounted for by genetic and environmental factors. "I do not share your belief that this will go far toward explaining the most interesting aspects of human life—those that have made us transcend our animality." He added that "scholars tend to overestimate this explanatory power of science and to underestimate the ability of humans (and probably of animals) to make choices and take decisions."[102] Dubos liked to defend this view by quoting the witticism of English author Samuel Johnson that "All theory is against freedom of the will, all experience for it."

Dubos was surprised to find it easy to write French in the lighthearted spirit of his youth, but found it difficult to translate this French into his mature English, so much so that many passages were entirely rewritten for the American edition. The book was slighted as unconvincing by reviewers in the United States; for example, geneticist Theodosius Dobzhansky found it "incomprehensible" that in dealing with an evolutionary problem there was "not a single mention in the book of natural selection."[103] In France, however, the book was the "antidote to pessimism" that Blanchard had predicted and brought Dubos immense public recognition. As a result, in his waning years, there were dozens of interviews in the French press in addition to a four-hour French television documentary and a book-length conversation based on his life produced by physician-scientist Jean-Paul Escande. A vast French public belatedly proclaimed him a "grand savant français."[104]

PURSUIT OF HAPPINESS. During America's bicentennial celebrations, Dubos took advantage of many happy occasions to reflect on his fifty years of adventures as an American. He used these occasions to link creative adaptations to such traditional American values as diversity, freedom, pursuit of happiness, and the right to a good environment.

For the opening celebration of New York City's observance of American Independence, Dubos was visibly moved by the honor of being the only speaker to address a crowd of thousands assembled at City Hall. Following a parade up Broadway, trumpet voluntaries, and patriotic songs, he spoke on "The Assets of Diversity" and depicted New York City as the place where people come "from all over the world to witness the exuberance of its urban life." He lauded it as "the experimental city par excellence" whose arenas of freedom, unlimited possibilities, and human diversity were essential to creativity. *The New York Times* published the full address at the top of the op-ed page that day.[105]

During the American Bicentennial in 1976, the United States Information Agency sent Dubos on an around-the-world lecture tour to Japan, Russia, Romania, Belgium, Italy, and France. His message concerned three aspects of creative adaptations to the future. He identified the human potential to act on behalf of situations that have not yet occurred, by anticipating consequences and pursuing new visions, as "the most impressive manifestation of freedom." He also defined two forms of pleasure necessary for health: *joie de vivre* as elemental biological well-being and enjoyment of being alive and *happiness* as delight in nonbiological activities (arts, sciences, ceremonies, adventures) that must be earned because it "depends upon acts of will and commonly involves struggles against hostile circumstances." This is why, he added, the Declaration of Independence wisely refers not to happiness itself but to each citizen's pursuit of it. Finally, he declared, "the right to a good environment is coming to be regarded as one of the fundamental inalienable rights." Five years later, just before he died, he was revising this last thought and envisioned a truly humanistic society that would "prize *joie de vivre* and happiness over the achievements of power and acquisition of wealth."[106]

Dubos also celebrated being an American in a very personal way by taking two years to write a fond tribute to his mentor, Oswald Avery, and to the vigorous, supple intellectual milieu of The Rockefeller Institute. Avery was remembered as a shy, puritanical, disciplined, cautious bacteriologist whose scientific achievements, save for the major discovery that DNA is the genetic material, were little known. The Institute was praised for having a "broad approach to understanding biological principles," where young men and women "could discover themselves." These tributes also reveal much about his own pursuit of happiness in this stimulating environment—a place where a foreign-born soil scientist, who knew nothing about medicine, could make many creative adaptations and succeed beyond the wildest dreams of a poor French boy.[107]

SELF-HEALING. During the late 1970s, "Doing better but feeling worse" was a common complaint from the general public and medical scientists. Adding to this charge, Dubos predicted medicine would not become fully scientific until it comes to grips with problems of social pathology, emotional disturbances, and the self-healing or resiliency mechanisms in a patient.[108] "I say that the important issue is understanding and promoting adaptation," meaning that the physician, after solving immediate medical problems, must teach patients how to meet and adapt to the associated emotional, social, and environmental issues.[109] "Rehabilitation implies active participation of the mind and of the body for a creative process of adaptive change," and this individual will can be motivated and guided by medical doctors.[110]

Somewhat surprisingly, especially to young physicians in the antibiotic era, he taught that healing does not necessarily imply a cure: "To heal may simply mean helping the patient to make curative adaptations to any kind of organic deficiencies . . . [and] to achieve a way of life that [is] tolerable and productive even in the presence of continuing disease."[111] This comment, of course, reflects his earlier definition of health as well as the advice given to him in the 1930s by his physician Homer Swift. This view continues to spread today among those who despair of the burdensome challenges of technological medicine that has everything to do with fixing symptoms, or "disease care," and little or nothing to do with true "health care."[112] The time may soon come, Dubos predicted in 1980, "when the emphasis in matters of health will have to shift from disabilities to functional ability and when the main question will be . . . to make the diseased or old person capable of functioning usefully and pleasantly for as long as biologically possible."[113]

Time and again during his life, Dubos had consciously applied self-healing, or self-renewal, to himself, but only in his final years did he speak openly about coping with his lifetime of disease. The reason for telling these stories was in part to acknowledge the help and advice from wise physicians, and, more profoundly, to offer himself as an example of the kind of self-renewal or health he advocated for others. He applied the example of a damaged tree to his experience: "when a branch breaks, the tree does not heal by replacing it, but by reorganizing its whole structure."[114] By this late age, he displayed an unspoken pride in living much longer than he ever expected as a sickly child and attributed this to consciously taking "decisions about the kind of behavior that I believe makes for a longer life." In particular, his permanent cardiac insufficiency and recurring ulcers had taught him to keep adapting to a progressively more contemplative, solitary lifestyle that at last began to limit his walking, climbing stairs, hiking woodland trails, and caring for his trees. "Planting trees has prolonged my life, for with it I project myself into the future."[115] By his own standards, the true measure of health is the extent to which we exert our potentials, however limited they may be, in anticipating and adapting to the future.[116]

Therefore, he prescribed, "Health is neither a gift, nor luck, it is an exercise of will."[117] He strongly believed that just as decisions affect normal activities, they can also influence an innate capacity to correct disabilities or pathologic processes. By engaging in "a creative adaptive process that requires choices and conscious participation by the whole organism" he affirmed it is possible to keep healthy, recover from disease, and live a full life.[118]

CREATIVE EVOLUTION, OR WILL THE FUTURE. Creative adaptations was also a genuine way for Dubos to express faith in an open, unlimited future of civilization. He had long distinguished human life, with its freedom of social evolution that is reversible and creative, from other animal life that remains prisoner of an irreversible biological evolution. Shortly before his death, he began to enhance this idea with creative adaptations, claiming that humans would improve on human nature by participating in an endless process of creative evolution.

In agreement with twentieth-century scientific philosophers Henri Bergson and Alfred North Whitehead, he envisioned a future reaching toward a balanced accommodation, in other words, an ever-present event. Bergson called this phenomenon "creative evolution," which "signifies both undivided continuity and creation."[119] Whitehead described evolution as "beneficent creativeness" in which "successful organisms modify their environment . . . to assist each other," so that when limitations disclose themselves, there is a "call for a renewed exercise of the creative imagination."[120] Dubos coined the more biological "creative adaptation" or, in his words, a "co-evolutionary" response of humans that involves a need to "will the future."

Along with many other scientists, he refuted the religious interpretation of evolution by another scientific philosopher, Pierre Teilhard de Chardin. Dubos was impatient with the "mystic and obscure scientific paraphernalia which shrouds Teilhard de Chardin's writings."[121] In a lecture prepared but not delivered for the centenary of his birth, he countered Chardin with the idea that the uniqueness of life cannot be accounted for entirely by physico-chemical or genetic mechanisms of evolution that run through all forms of life like a continuous thread. Even if Chardin had lived to restate evolution in terms of DNA molecules, Dubos claimed, this "would probably not have had a profound influence on his philosophy of evolution," although some modern discoveries "*not* related to genes might have." With this, he pointed to evidence accumulating from the new molecular biology that most fundamental innovations in life forms occurred very soon after the "origin" of life itself. Evolution has proceeded, he observed, "not by creating biochemical novelties but by rearranging over and over again the same fundamental elements that still constitute today the metabolic and structural aspects of Life."[122]

The Dubos contribution to creative evolution was his observation that because human beings exercise free will, they have acted throughout history and can act in the future to change their environments and their lives. "The only trouble with the [traditional] theory of evolution is that it camouflages all questions behind the appealing metaphor—evolution of the Universe, evolution of Life, evolution of Mind, evolution of Culture." As a biologist, he acknowledged that the higher an organism is on the evolutionary scale, the more exacting its interdependence and integration in a complex ecosystem. This trend, implied by Chardin's progression toward Omega Point where all become one, he said, is "not a noble thought but a denial of the universal evolutionary trends which make for greater and greater diversity."[123] As an ecologist, he pointed to a simultaneous and opposite trend toward "freedom and independence that also intensifies as one ascends the evolutionary scale," a trend he said was not being addressed by other evolutionary thinkers. He ended on an understated, quiet note: "Darwinian evolution postulates that, in the competition for survival, the reward goes to the fittest. But the meek may also inherit the earth through the creativeness of their adaptations and associations."[124]

Dubos thus concluded his search for the meaning of health as a creative adaptation that will initiate more sensible and meaningful ways of life.

Almost from his youth, René Dubos was a despairing optimist, a practical skeptic, who balanced despair with hope. The personal, familial, and national tragedies during his formative years in France taught him that people and societies can recover from many ordeals and then move on to become successful. As a scientist who spanned the century between the Age of Germs and the Age of Genomes, he often stood alone on ideas and principles, many of which continue to grow in importance. During experimental studies, he discovered time and again the potentials and resiliency of natural systems, and this science fueled his irritation and pessimism about the mismanagement of modern life and the degradation of mankind. While continuing to despair, he turned to teaching that the finest forms of civilization are those that use science and technology to serve basic biological needs and generate values that enhance the quality of life. To the biologist, "inertia is the only mortal danger."[125] To the humanist, "optimism is essential for action and constitutes the only attitude compatible with sanity."[126] Dubos had an unshakable faith in human beings, the true starting point for change, because they can demand what is good for life and ultimately will the future.

Near the end of his life, Dubos was asked whether as a biologist he had come to terms with the inevitability of his own death. He replied that it did not preoccupy him, saying "I suspect that eventually a certain lassitude, or perhaps a feeling of satiety somewhat like one has after a full meal, will overtake me, a feeling that says, in essence, 'I've had enough' . . . a built-in trait that prevents obsession with the idea of death." Citing bacteriologist Élie Metchnikoff, he added, "At the end of a long day, when you have been fully occupied, the time comes to rest."[127]

René Dubos died on his eighty-first birthday, 20 February 1982, in New York Hospital of pancreatic cancer after a three-week illness. As he long wished, he died while listening to cherished Gregorian chants and French church bells on recordings that Jean brought from home that morning. It was the beginning of Lent, spring, his favorite time to walk among his trees and to witness once more the renewal of nature. He was about to begin the autobiography whose title he had chosen years before, *Le printemps m'appelle*, Spring is Calling Me.

Envoi

By the end of his life, René Dubos saw the environmental movement, whose philosophical underpinnings he had helped shape, grow from a small fringe element to a major political and cultural force. Several of his prophetic warnings and fears have become realities, and their problems reach into our daily lives. Some of his hopes have also been justified, for predicted eco-catastrophes have not occurred, and the year 2000 found itself with new growth, ideas, innovations, and possibilities.

What he brought to this movement was an ecological conscience that sustained an immensely successful scientific career. Scientifically defined, ecology is simply the study of relationships between organisms and their environment. His broader vision incorporates the human presence and introduces an ethical component. Judgments based on knowledge need to be value conscious and value oriented, because choices and actions affect the *quality* of these relationships. Maintaining human health and the health of the earth are ecological imperatives and ones that must be evaluated in the light of human needs, tastes, and aspirations.

The environmental movement that continues to shape countless disciplines is pervaded and enlarged by Dubos' perceptions that we must stop conquering natural ecosystems, treat them with respect, and live more thoughtfully among them. For these concerns, his books and essays may be more relevant now than when they were written. His visions reach beyond conventional gloom with heartening beliefs that diversity, innovation, and knowledge are ways to preserve, enrich, and create healthy environments. The essential biotic view of health, to rephrase his grand motto, is think ecologically, act ethically.

As a laboratory scientist, he deferred to experimental evidence yet took away larger messages. As a lifelong student of medicine and history, he knew that changing environments may have as much or more to do with susceptibility or resistance to disease than anything in a physician's armamentarium. As an ecologist, he raised awareness of human dependence on the vast ecosystems surrounding us, from agriculture, medicine, technology, energy, and architecture to working and living habitats. As a teacher, he created simple mottos to convey complex abstractions. The prescient "mirage of health," "improving on nature," "trend is not destiny," "creative adaptations," and "will the future" speak directly to ethics and behavior. To whichever ecosystem he applied his thinking, his integrative wisdom focused on seeking health. Above all, as a good doctor, he practiced ecology as a healing art.

Acknowledgments

After celebrating the centenary of René Dubos' birth at a reminiscence-filled lunch in a French bistro in Manhattan, I returned home to get the details of an event from my diary. By chance, I saw an entry from August 1978 where I noted that Dubos had remarked that perhaps I would be the one to write his biography. The entry tells two tales. The first affirmed our long professional relationship during which I had completed my doctorate in comparative literature at Columbia University while working with him on a wide range of projects. The other acknowledged that he thought a biography would one day be merited.

Dubos had often talked about writing an autobiography, and we occasionally talked about what would be in it, but nothing was written. His final book, *Celebrations of Life*, included a few details from the early years but it was far from an autobiography. While apologizing to his editor for not achieving this goal, he introduced her to me as the one who knew the most about him. This surprised me because the sparse anecdotes he published were the ones he revealed on the rare occasions when he talked about himself. To the end, he remained a private person.

So it fell to his wife, Jean, after his death in 1982, to help realize the elusive goal of a biography. This book is the first full-length story of his life and reflects her determination that there should be an authoritative public record of the whole of her husband's life and work. In the six years before her death in 1988, she took an active role in keeping his books in print. She also took a keen interest in supporting my preparation of a complete, annotated bibliography. Neither of us knew at the time that the final list of two dozen books and nearly a thousand essays and reviews was three times longer than any record he kept for himself. I benefitted greatly from Jean's perceptive comments about their forty years together and reading her few personal papers. That the bibliography became the principal source and starting point for the biography was due to the enthusiasm and sustained support of Jean, the Richard Lounsbery Foundation, Alan McHenry, Frederick Seitz, and Ralph Steinman.

There were real complications in expanding the rich library of ideas from Dubos' mind, as found in his prolific writings, into a full-life portrait of the man who wrote them. An initial survey of the materials available was discouraging. Before 1970 Dubos discarded his laboratory notebooks, correspondence, and personal papers. He purged these materials regularly, and not until the Rockefeller University Archives was established did he save selected correspondence and manuscripts. The documents from the last decade of his life allow a comparatively better description of the environmental years than the previous seventy. Even after making extensive searches for letters with presumed correspondents and interviews with those who knew him before 1950, knowledge of the first two-thirds of his life remains tantalizingly fragmentary.

A further obstacle to accurate biographical information comes from four documents in which he described his early life. In addition to *Celebrations of Life*, the other sources are interviews: an oral history taken for Columbia University by Saul Benison (1955–1956), an anonymously rewritten and slightly updated version of the oral history for *Medical World News* (1975), and a dialogue with Jean-Paul Escande in *Quest* (1980). My background research revealed that these colorful yet highly selective reminiscences contain discrepancies, inconsistencies, conflated events, and many errors. Lack of documentation was particularly frustrating to Benison, who complained that Dubos "did not make available to me any of his private correspondence." This was, of course, because there was none. I further discovered that little or no fact checking was done by Dubos, Benison, or Escande of names, dates, or events mentioned in these interviews.

Unfortunately, these undocumented and unverified sources have served as the basis for many published profiles of Dubos. He contributed to this problem by lending the oral history so he could avoid answering the same questions again and again. As a result, the profile writers who did not verify the facts they reported have perpetuated errors about important historical facts and events. For example, although Dubos claimed he stopped looking for antibiotics after gramicidin, many documents show that he spent another ten years searching for them. In addition, since the profiles focus on events before 1955, the writers have assumed that Dubos' fame rests on his discovery of antibiotics and dismiss the remaining, most fruitful three decades of his life. In contrast, as this biography presents for the first time, Dubos did pioneering research in fields other than microbiology for nearly two more decades, wrote two dozen books, and became famous for many later discoveries.

These obstacles pose a serious challenge to a biographer. On one hand, the value of his reminiscences, oral history, and interviews lies in the spirit of what he was trying to convey. His opinions and perceptions of environments, people, and their influences are trustworthy guides to his feelings at the time they were expressed. On the other hand, a closer reading in this lore of memory reveals valuable clues about his behavior. There are disturbing silences on some topics and unaccounted years in his life, possibly due to limited questions from interviewers, but just as likely due to choices made by Dubos to create a self-image. Unknown to everyone who has relied solely on the oral history is that what he chose to remember in 1955 was colored by illness and depression. The real questions of concern to a biographer are what he was trying to avoid or to achieve at that time and whether these early sentiments and self-image changed. These and many other questions have now been examined in the context of his whole life.

Consequently, this biography is based on newly researched primary sources that have added surprising details and important links to his personality. Many stories are told for the first time. Some of Dubos' own anecdotes have been amplified or modified as they were verified by other sources. Other stories expand, change, or revise what has been previously published. Even so, gaps remain, some that can be ascribed to his primarily sequestered lifestyle and some that are odd and troublesome to the tidy habits of a biographer. The activities described about his later years, in contrast, represent my interactions with René, Jean, their friends, colleagues, and relatives.

Only after finishing this biography did I realize why Dubos had remarked I was the one who knew the most about him. He had shared with me more of his innermost thoughts than anyone except his two wives. What I learned were not his everyday preoccupations or gossipy confessions. I was fortunate to experience his endless ideas. In many particulars, I found him a literary heir of sixteenth-century French skeptic and essayist Michel de Montaigne, whose hopes, fears, concerns, and beliefs expressed in writing divulged his character and changing moods. Like Montaigne, the essays of Dubos constitute an engaging self-portrait.

The impenetrable reserve that once seemed hidden beneath this portrait also became less mysterious while I was researching his life. During the difficult early years, it was clear that he exercised immense self-discipline to overcome fragile health and live a long life. This reserve subsided during his final years when he began to talk about his illnesses and revealed how he too was a member of this flawed but wonderful human species. A more concealed aspect of his personality was the drive to succeed on his own powers. Despite setbacks and disappointments, he displayed an uncanny ability to adapt and to anticipate new circumstances. This behavior led to many successes after the antibiotic discovery. In the end, his childhood dream to someday be a champion often explained his motivations. His long adventure with life might be best expressed in a motto by Henri Pélissier, the most famous French bicycling champion during Dubos' youth. "To win one race is nothing, but to win several more, that counts."

Many personal stories and details in this biography were gleaned with help from numerous individuals and sources.

I am especially indebted to Rollin D. Hotchkiss, a major figure in the gramicidin discovery. He knew and worked alongside Dubos longer than anyone except Jean. He generously allowed me to draw on his recollections of their life in Oswald Avery's laboratory. Rollin read an early manuscript and made constructive comments, all the while providing enjoyable and profitable hours of intense conversations as well as long, thoughtful letters.

The Rockefeller Institute for Medical Research (now The Rockefeller University) was the center of Dubos' professional life. I would like to express my gratitude to colleagues in his laboratory and members of the University staff who have given much help and advice, especially the laboratory heads who were successors to the Dubos laboratory: first James G. Hirsch, then Zanvil A. Cohn, and now Ralph M. Steinman. An early manuscript was read by Ralph, who made astute suggestions for its improvement. At several stages in the research and writing, Frederick Seitz and Joshua Lederberg, former presidents of the University, helped me greatly and directed me to pertinent background information. I also profited from extended discussions with Cynthia Pierce-Chase, Merrill W. Chase, and Bernard D. Davis, members of the tuberculosis laboratory during the 1940s.

For valuable perspectives on launching the antibiotic era, I am indebted to Norman G. Heatley, Edward P. Abraham, Theodore Woodward, George Mackaness, along with Rollin Hotchkiss and Bernard Davis, who shared personal accounts of their discoveries during the 1940s. I am particularly grateful for the hospitality of Norman Heatley and charming discussions and excursions with him at the Sir William Dunn School at Oxford during the Penicillin 50 celebration.

During my research I have appreciated the exchanges with many individuals who offered suggestions and shared memories of people and events.

For advice and insight into the lore of the pneumococcus, streptococcus, tubercle bacillus, staphylococcus, and multiple other microbes, I thank Maclyn McCarty, Walsh McDermott, Marcus Horwitz, Carl Nathan, Michel Rabinovitch, Arturo Zychlinsky, and John Zabriskie. For candid descriptions of Institute life, my stories have been enriched by Vincent Dole, Ian Maclean Smith, Walther Goebel, James S. Murphy, Keith Porter, David Lyons, Mabel Bright, Egilde Seravalli, and Lila Magie. For leads to the young years and French connections, I thank André Draghi, Robert Fauve, Byron Waksman, Simone Bramerel, François Léry, François Marchon, and Jean-Paul Escande. For aspects of history of science, medicine, and the environment, I appreciate the observations of Paul Shepard, Lawrence K. Altman, Saul Benison, Joseph Fruton, Curt Meine, Alexander Bearn, Robert Coghill, William K. Reilly, and Lewis Thomas. For fathoming some New York City and global encounters of René and Jean, I am grateful for the perspectives of André Cournand, David Rogers, Barbara Ward, Francis Trudeau, Yehudi Menuhin, and James Parks Morton.

Materials from family members and close friends added valuable details. René's sister Marie Madeleine, brother Francis, and nephew Georges-André have given interviews, photographs, and information in letters and shared early correspondence of René with his mother. Jean's niece Jennifer Porter and sister-in-law Lucille Porter provided photographs and family history during personal interviews. Henriette Noufflard Guy-Loë and Frank Fenner gave me copies of their private correspondence with René and Jean during the 1950s. Keith Porter graciously shared his reminiscences and voluminous letters concerning antimycobacterial research at the Ray Brook sanatorium.

The Rockefeller Archive Center was very helpful, especially Renee Mastrocco. Other archivists made notable contributions to the story. M. B. Clabaut at the International Society of Soil Science in Wageningen, the Netherlands, helped with information on soil congresses in 1924 and 1927. Jeffrey Karr found correspondence related to the Rutgers years in the archives of the American Society for Microbiology. Beth Harris at Hollins University, Roanoke, Virginia, and Edburne R. Hare at The Masters School, Dobbs Ferry, New York, found materials related to Marie Louise Bonnet. Daniel Barbiero and Janice Goldblum at the National Academy of Sciences, Washington, D.C., Arthur Anderson at Fort Detrick, Maryland, and Marjorie Ciarlante and Wilbert Mahoney at the National Archives, College Park, Maryland, provided valuable leads and assistance with War Research Service, OSRD, and Chemical Warfare Service materials. Richard J. Wolfe at the Francis A. Countway Library of Medicine, Harvard Medical School, located documents related to the Harvard years. Beth Carroll-Horrocks and Scott DeHaven aided my research in the Peyton Rous Papers at the American Philosophical Society in Philadelphia. Anne Sardon at the Ministère de l'Environnement, Annick Perrot and Robert Fauve at Institut Pasteur, and Emmanuel LeRoy Ladurie, director of La Bibliothèque Nationale, expedited my research in France.

As the biography neared completion, I benefitted greatly from the close reading of the manuscript by Alexander Tomasz, Thomas D. Brock, Douglas E. Eveleigh, and William R. Jacobs, Jr.

Judith Schwartz, my first editor, deserves special thanks. She asked me to write my first article on Dubos and during this project offered many wise observations.

Lastly, I am pleased to thank my husband Lawrence Moberg, who has worked with me throughout many years of research and writing, for his tough questions, constant help, and cheerful encouragement.

Because these people have helped me does not mean they approve or are responsible for what I have written. That burden, as well as errors and omissions, is mine.

Carol Larsen Moberg
20 February 2005

Appendix I
Events of René Dubos' Life

1901	Birth on 20 February at Saint-Brice-sous-Forêt, France. Parents Georges Alexandre Dubos and Madeleine Adéline De Bloedt. Two siblings, sister Marie Madeleine (1903–1991) and brother Francis (1907–2003).
1903	Family moved to Hénonville.
1909	First attack of rheumatic fever; bedridden for nearly two years.
1914	Family moved to Paris.
1914–1918	Awarded a scholarship for secondary education at Collège Chaptal, Paris.
1919	Third attack of rheumatic fever.
	Death of father, shortly after his return from military service.
1921	Degree of *Ingénieur*, Institut National Agronomique, Paris.
1922	Degree of *Ingénieur*, Institut National d'Agronomie Coloniale, Nogent-sur-Marne.
1923	Associate Editor, *International Review of the Science and Practice of Agriculture* at the International Institute of Agriculture in Rome.
1924	Immigrated to the United States.
1926	Master of Science, Rutgers University, New Brunswick, New Jersey.

1927	Doctor of Philosophy, Rutgers University, under Selman Waksman.
	Appointment as fellow, Department of the Hospital, The Rockefeller Institute for Medical Research (RIMR), New York City, in the laboratory of Oswald T. Avery.
1928	Assistant, RIMR.
	Begins studies on capsular polysaccharide of the pneumococcus.
1930	Associate, RIMR.
	Begins studies on bacterial enzymes.
1933	Fourth attack of rheumatic fever.
1934	Marriage to Marie Louise Bonnet (1898–1942) on 23 March at the Church of Notre Dame, New York City.
1938	Associate member, RIMR.
	Studies on bactericidal agents from soil microbes.
	Naturalized as a United States citizen on 5 December.
1939	Reports discovery of the antibiotics tyrothricin, tyrocidine, and gramicidin.
1940	John Phillips Memorial Award (the first of more than thirty major prizes received during his lifetime).
1941	Member, RIMR.
	Elected to the National Academy of Sciences.
	Honorary degree from University of Rochester, New York (the first of forty-one degrees received from other colleges and universities during his lifetime).
1942	Death of Marie Louise Dubos on 24 April; burial at Gate of Heaven Cemetery, Hawthorne, New York.
	Resignation from RIMR.
1942–1944	Appointment as George Fabyan Professor of Comparative Pathology and Tropical Medicine, School of Public Health, Harvard University, Boston, Massachusetts.
1944	Resignation from Harvard and reappointment as member, RIMR.
	Establishes laboratory to study tuberculosis.

1945	Publication of his first book, *The Bacterial Cell.*
1946	Marriage to Letha Jean Porter (1918–1988) on 16 October at The Church of the Epiphany, New York City.
	Purchases abandoned farm in Garrison, New York.
	Appointed an editor of *The Journal of Experimental Medicine.*
1948	Albert Lasker Medical Research Award.
1950	Publication of *Louis Pasteur: Free Lance of Science.*
	Begins studies on infection versus disease and host-parasite relationships.
1951	Trudeau Medal from the National Tuberculosis Association.
1952	President, Society of American Bacteriologists (now the American Society for Microbiology).
	President, The Harvey Society.
	Publication, with Jean Dubos, of *The White Plague.*
1952–1962	Technical advisor on standardizing BCG vaccine, Centre International de l'Enfance, Boulogne, France.
1954	Elected to the American Philosophical Society.
	Publication of *Biochemical Determinants of Microbial Diseases.*
	Death of mother on 24 April.
1955	Death of Oswald Avery on 20 February.
1957–1971	Professor and member, The Rockefeller University.
1959	Publication of *Mirage of Health.*
1960	Robert Koch Medal, Berlin, Germany.
1961	Begins studies on environmental biomedicine.
1962	Publication of *The Torch of Life.*
1965	Publication of *Man Adapting.*
1966	Arches of Science Award from the Pacific Science Center, Seattle, Washington.

1968	Publication of *So Human an Animal.*
1969	Pulitzer Prize for Nonfiction for *So Human an Animal.*
1969–1975	Appointed by President Richard Nixon to the Citizens' Advisory Committee on Environmental Quality.
1969	Elected Benjamin Franklin Fellow of Royal Society of Arts, London, England.
1970	Begins "The Despairing Optimist" column for *The American Scholar,* which continues until 1980.
1971	Professor emeritus, The Rockefeller University. Long convalescence from subacute bacterial endocarditis.
1972	Publication of *A God Within* and, with Barbara Ward, of *Only One Earth.* Prix de l'Institut de la Vie, Paris, France.
1974	Simultaneous publication of *Choisir d'être humain* and *Beast or Angel? Choices that Make Us Human.*
1975	Cullum Geographical Medal from the American Geographical Society.
1976	Tyler Prize. Publication of *The Professor, the Institute, and DNA.*
1979	Elected to the American Academy and Institute of Arts and Letters. Wilder Penfield Award, Ottawa, Canada.
1980	Publication of *The Wooing of Earth.*
1981	Publication of *Celebrations of Life.* Prix Sully–Olivier de Serres for *The Wooing of Earth,* Académie Française, Paris, France.
1982	Death on 20 February in New York Hospital, New York City. Ashes buried at St. Philips in the Highlands, Garrison, New York.

APPENDIX II
Books, Monographs, and Pamphlets by René Dubos
(in chronological order)

The Bacterial Cell in its Relation to Problems of Virulence, Immunity and Chemotherapy (Cambridge: Harvard University Press, 1945). (Cited as *The Bacterial Cell*)

Editor, *Bacterial and Mycotic Infections of Man* (Philadelphia: Lippincott, first edition, 1948; second edition, 1952; third edition, 1958; and fourth edition, with James G. Hirsch, 1965).

Louis Pasteur: Free Lance of Science (Boston: Little, Brown and Company, 1950). (Cited as *Louis Pasteur*)

With Jean Dubos, *The White Plague: Tuberculosis, Man, and Society* (Boston: Little Brown and Company, 1952). (Cited as *The White Plague*)

Biochemical Determinants of Microbial Diseases (Cambridge: Harvard University Press, 1954). (Cited as *Biochemical Determinants*)

Mirage of Health: Utopias, Progress, and Biological Change (New York: Harper & Brothers, 1959). (Cited as *Mirage of Health*)

Pasteur and Modern Science (Garden City, New York: Doubleday Anchor Books, 1960).

The Dreams of Reason: Science and Utopias (New York: Columbia University Press, 1961). (Cited as *Dreams of Reason*)

Evolution of Concepts in the Prevention of Tuberculosis, The Baker Lecture (Ann Arbor: University of Michigan School of Public Health and Michigan Tuberculosis Association, 1961).

The Torch of Life: Continuity in Living Experience (New York: Simon and Schuster, 1962). (Cited as *The Torch of Life*)

The Unseen World (New York: The Rockefeller Institute Press and Oxford University Press, 1962).

The Cultural Roots and the Social Fruits of Science, Condon Lectures (Eugene, Oregon: Oregon State System of Higher Education, 1963).

With Maya Pines and the Editors of Life, *Health and Disease* (New York: Time, Inc., 1965).

Man Adapting (New Haven, Connecticut: Yale University Press, 1965).

Man and his Environment: Biomedical Knowledge and Social Action (Washington, D.C.: Pan American Health Organization Publication 131, 1966).

Man, Medicine, and Environment, A Britannica Perspective (New York: Frederick A. Praeger, 1968).

So Human an Animal (New York: Charles Scribner's Sons, 1968).

A Theology of the Earth (Washington, D.C.: Smithsonian Institution, 1969).

Reason Awake: Science for Man (New York: Columbia University Press, 1970). (Cited as *Reason Awake*)

The Genius of the Place, Horace M. Albright Conservation Lecture (Berkeley: University of California School of Forestry and Conservation, 1970). (Cited as *Genius of the Place*)

A God Within (New York: Charles Scribner's Sons, 1972).

With Barbara Ward, *Only One Earth: The Care and Maintenance of a Small Planet* (New York: W.W. Norton, 1972). (Cited as *Only One Earth*)

Humanizing the Earth, The 1972 B.Y. Morrison Memorial Lecture (Washington, D.C.: Agricultural Research Service, United States Department of Agriculture, 1973).

From Nature to Resources or Does Nature Really Know Best, Joseph Wunsch Lecture (Haifa, Israel: Technion, Israel Institute of Technology, 1973).

Choisir d'être humain (Paris: Denoël) published simultaneously in English as *Beast or Angel? Choices That Make Us Human* (New York: Charles Scribner's Sons, 1974). (Cited as *Beast or Angel*)

Of Human Diversity, The 1972 Heinz Werner Lecture (Barre, Massachusetts: Clark University Press with Barre Publishers, 1974).

The Professor, the Institute, and DNA: Oswald T. Avery, His Life and Scientific Achievements (New York: The Rockefeller University Press, 1976). (Cited as *The Professor*)

The Resilience of Ecosystems (Rome: Accademia Nazionale dei Lincei, 1977).

The Resilience of Ecosystems: An Ecological View of Environmental Restoration, Reuben G. Gustavson Lecture (Boulder: Colorado Associated University Press, 1978).

Human Development and the Social Environment, Wilder Penfield Award Lecture (Ottawa, Canada: The Vanier Institute of the Family, 1979).

The Wooing of Earth (New York: Charles Scribner's Sons, 1980). (Cited as *Wooing of Earth*)

Celebrations of Life (New York: McGraw-Hill Book Company, 1981).

Note: In addition to these major works, René Dubos published 236 scientific research articles, 295 essays and lectures, and 260 short items (i.e., reviews, forewords, op-ed pieces). The collection of his papers in the Rockefeller University Archives also includes 157 unpublished manuscripts of essays and lectures that range in length from two to 127 pages. Many of these articles and manuscripts are cited in full in the endnote references that follow.

Notes

For frequently cited works in the Notes, some short titles and abbreviations have been used.

ART	*American Review of Tuberculosis* and its subsequent title *American Review of Tuberculosis and Pulmonary Diseases*
DO	"The Despairing Optimist," a column appearing in *The American Scholar* from 1970 to 1980
JEM	*The Journal of Experimental Medicine*
NAS	Archives, National Academy of Sciences, Washington, D.C.
OH	Oral History. "Reminiscences of Dr. R. Dubos: A Personal Commentary on a Life in Science," an oral history memoir taken October 1955 and August 1956 by Saul Benison in the Columbia University Oral History Research Office Collection. Citations from this work are used courtesy of the Oral History Research Office. The copy of the OH used by the author is René Dubos' original transcript, which has been annotated over the years by several colleagues, Jean Dubos, and the author. The page numbers given first in the references are to this copy; the numbers in parentheses are to the pages in the Columbia University document that intersperses Dubos' published scientific articles throughout the transcript.

OSRD	The records and correspondence of the Office of Scientific Research and Development are in Record Group 227 of The National Archives, College Park, Maryland. Individual citations to this material include the box number and title.
RD	René Dubos
RD Family Papers	Private correspondence belonging to the Dubos Family, Paris, France. All translations from the original French are my own.
RP	Robert Potet
RP Letters	The correspondence between RD and RP is found in the RD Papers, Box 128, Folder 28. All translations from the original French are my own.
RS	Robert Starkey
RS Papers	Robert Starkey Papers, Folder 15.1, 13-II BM, The American Society for Microbiology Archives, Albin O. Kuhn Library, University of Maryland Baltimore County, Baltimore, Maryland.
RUA	The Rockefeller University Archives, Rockefeller Archive Center, Sleepy Hollow, New York. Two frequently cited RUA collections include:

BSD	*Annual Scientific Reports to the Board of Scientific Directors of the Rockefeller Institute for Medical Research*, Record Group 439. This unpublished collection comprises 42 volumes, from 1901 to 1954, and contains detailed reports written by each member about research in the laboratory.
RD Papers	René Dubos Papers, Record Group 450 D851.

Chapter 1 Orchestral Relationships and Soil Microbes

1. Facts about the early years are based on documents, including school transcripts, registers, certificates, photographs, and letters in the author's papers, interviews by the author with René's brother Francis, sister Madeleine, nephews Georges-André and Jacques, and notes taken after lengthy conversations with both René and Jean Dubos. Many facts are at variance from the Oral History and these published versions: "Medicine's Living History: Dr. René Dubos," *Medical World News* 16 (5 May 1975): 77–87; Jean-Paul Escande and RD, *Quest: Reflections on Medicine, Science and Humanity*, translated by Patricia Ranum (New York: Harcourt Brace Jovanovich, 1980); and RD, *Celebrations of Life*.

Concerning the name Dubos, it "can be found all over ancient Picardy (it is the picard form of Dubois), and it is used especially in Somme and Oise but appears also in the nearby departments . . . the name Dubos is frequent in Normandy where it can be written also Dubosc and even Duboscq or, more simply, Dubois." G. Dubos-Cantais to RD, 19 October 1981, RD Papers.

RD used and published variant spellings of his mother's name, from Adéline Madeleine DuBlöedt in his 1927 application to The Rockefeller Institute, to De Bloëdt and Adeline De Bloedt in *Celebrations of Life* (1981). On René's baptismal certificate dated 26 May 1901, she gave her name as Madeleine Adéline De Bloedt and that of her brother as Jules De Bloedt, René's godfather. For consistency, the variant spelling of their name in this certificate is used throughout. From 1914 to 1918, Jules De Bloedt lived in Paris with Adéline, RD, Madeleine, and Francis while working in a munitions factory. RD to Gabriel Renaudin (a great-grandson of Pierre De Bloedt), 26 February 1981, RD Papers.

2. RD, "A Celebration of Life," in *The World of René Dubos: A Collection from His Writings*, edited by Gerard Piel and Osborn Segerberg, Jr. (New York: Henry Holt and Company, 1990), pp. 47–50. This selection includes material not in the original publication: *United Magazine*, April 1982, pp. 9, 102.

3. RD to Barbara Tuchman, 9 February 1981, RD Papers.

4. Francis Dubos provided information about Georges Alexandre's military service, circumstances of his death, and closing of the butcher shop. The *épicerie* is described in a letter from RD to Madeleine Héry, 9 March 1977, RD Papers.

5. Richard Kostelanetz, "The Five Careers of René Dubos," *Michigan Quarterly Review* 19 (Spring 1980): 194–202.

6. RD, "The Relative Importance of the Carbon Dioxide of the Soil and of the Atmosphere in Plant Growth," *International Review of the Science and Practice of Agriculture* 2 (October–December 1924): 864–867.

7. William Bryant Logan, "Hans Jenny at the Pygmy Forest," *Orion* 11 (1992): 18–29.

8. This section relies notably on these sources: Edward J. Russell, *The Micro-organisms of the Soil* (London: Longmans, Green, 1923); Selman A. Waksman, *Principles of Soil Microbiology* (Baltimore: Williams & Wilkins Company, 1927); Sergei Winogradsky, "Études sur la microbiologie du sol. IV. Sur la dégradation de la cellulose dans le sol," *Annales de l'Institut Pasteur* 43 (1929): 549–633; and Selman A. Waksman, *Sergei N. Winogradsky: His Life and Work* (New Brunswick: Rutgers University Press, 1953).

9. RD, "Does Man Have a Nature?" *The Rockefeller University Review* 4 (1966): 1–4.

10. Sergei Winogradsky, "La méthode directe dans l'étude microbiologique du sol," *Chimie et Industrie* 11 (February 1924): 215–222.

11. OH, Part I, p. 14 (14).

12. Sergei Winogradsky, "Mémoire," *Annales de l'Institut Pasteur* 4 (1890): 211–31, quoted in Hubert A. Lechevalier and Morris Solotorovsky, *Three Centuries of Microbiology* (New York: McGraw-Hill Book Company, 1965), p. 266.

13. Winogradsky, "La méthode directe."

14. Thomas D. Brock and Michael T. Madigan, "Winogradsky's Legacy" and "Enrichment and Isolation Methods," *Biology of Microorganisms*, sixth edition (Englewood Cliffs, N.J.: Prentice Hall, 1991), pp. 581, 614–617.

15. RD, Two handwritten pages on Sergei Winogradsky, n.d., author's papers.

16. Selman A. Waksman, "Soil Microbiology in 1924: An Attempt at an Analysis and a Synthesis," *Soil Science* 19 (1925): 201–246.

17. Selman A. Waksman, "Dr. René J. Dubos—A Tribute," *Journal of the American Medical Association* 174 (1960): 111–113.

18. Biographical details about Waksman can be found in Rollin D. Hotchkiss, "Selman Abraham Waksman, 1888–1973," *Biographical Memoirs* 83 (2003): 320–343; Hubert A. Lechevalier, "The Waksman Institute of Microbiology (1954–1984)," *The Journal of the Rutgers University Libraries* 50 (1988): 20–45; and Selman A. Waksman, *My Life with Microbes* (New York: Simon & Schuster, 1954).

19. A discussion of RD's place in the nuanced field of *human ecology* appears in chapter 5 where the term is described as a watchword of the environmental movement.

20. OH, Part I, pp. 22, 20 (22, 20).

21. OH, Part I, p. 21 (21).

22. Waksman, "Dubos—A Tribute."

23. Selman A. Waksman and RD, "Microbiological Analysis of Soils as an Index of Soil Fertility. X. The Catalytic Power of the Soil," *Soil Science* 22 (1926): 407–420.

24. Quoted in Waksman, *Sergei N. Winogradsky,* p. 90.

25. RD, "A Study of the Decomposition of Hydrogen Peroxide by Soil and its Possible Use as an Index of Soil Fertility" (M.S. dissertation, Rutgers University, April 1926), pp. 5, 46, 1.

26. RD to Tom McCall, 22 April 1981, RD Papers; and RD, "Does Man Have a Nature?"

27. Russell, *Micro-organisms of the Soil,* p. 28.

28. V.L. Omeliansky, "Zur Frage der Zellulose Gärung," *Centralblatt für Bakteriologie,* 2 Abt., 36 (1913): 472–473.

29. K.F. Kellerman, I.G. McBeth, F.M. Scales, and N.R. Smith, "Identification and Classification of Cellulose Dissolving Bacteria," *Centralblatt für Bakteriologie*, 2 Abt., 39 (1913): 502–522.

30. Selman A. Waksman and H. Heukelekian, "Microbiological Analysis of Soil as an Index of Soil Fertility. VIII. Decomposition of Cellulose," *Soil Science* 17 (1924): 275–291; Charles E. Skinner, "Organisms Responsible for Cellulose Decomposition in Soil" (Ph.D. dissertation, Rutgers University, 1925); and Selman A. Waksman and Charles E. Skinner, "Microorganisms Concerned in the Decomposition of the Soil," *Journal of Bacteriology* 12 (1926): 57–84.

31. RD, "Concerning the Nature of the Flora Active in the Decomposition of Cellulose in the Soil, with Special Regard to the Activity of Bacteria in this Process" (Ph.D. dissertation, Rutgers University, April 1927), pp. 6, 58.

32. Sergei Winogradsky, "Sur la décomposition de la cellulose dans le sol," *Comptes Rendus des Séances de l'Académie des Sciences* 183 (1926): 691–694, quoted in RD, "Concerning the Nature of the Flora," p. 44.

33. Waksman, "Dubos—A Tribute."

34. RD, "Concerning the Nature of the Flora," pp. 102, 98.

35. Selman A. Waksman and RD, "Sur la nature des organismes qui décomposent la cellulose dans les terres arables," *Comptes Rendus des Séances de l'Académie des Sciences* 185 (1927): 1226–1228.

36. RD, "Influence of Environmental Conditions on the Activities of Cellulose Decomposing Organisms in the Soil," *Ecology* 9 (1928): 12–27.

37. RD, "The Decomposition of Cellulose by Aerobic Bacteria," *Journal of Bacteriology* 15 (1928): 223–234.

38. Winogradsky, "Études sur la microbiologie," pp. 554, 630.

Chapter 2 Domesticating Microbes

1. Frederick T. Gates, Memorandum to Starr J. Murphy, 31 December 1915, quoted in George W. Corner, *A History of The Rockefeller Institute, 1901–1953: Origins and Growth* (New York: The Rockefeller Institute Press, 1964), pp. 575–584.

2. RD, "Chemistry in Medical Research," *The Professor,* pp. 35–46.

3. OH, Part I, p. 37 (37).

4. RD Papers, Box 2, Folder 1. This correspondence shows RD's acceptance of a Rockefeller appointment before leaving on a transcontinental train tour with soil scientists, thus contradicting his other reminiscences of receiving a job offer by telegram from the Institute while he was in Fargo, North Dakota, on this tour.

5. Herbert S. Gasser, "Report on the Activities and Problems of The Rockefeller Institute," *BSD* 28 (3 October 1939): 1–86.

6. Edward J. Russell, *The Land Called Me: An Autobiography* (London: Allen & Unwin, 1956), p. 161.

7. RD Papers, Box 18, Folder 14.

8. "Medicine's Living History: Dr. René Dubos," *Medical World News* 16 (5 May 1975): 77–87.

9. Oswald T. Avery, H.T. Chickering, Rufus Cole, and A.R. Dochez, *Acute Lobar Pneumonia: Prevention and Serum Treatment* (New York: The Rockefeller Institute for Medical Research Monograph Number 7, 1917); and Simon Flexner, "Twenty-five Year Sketch of Rockefeller Institute," *BSD* 18 (1929–1930): 78–150.

10. RD, *The Professor*, pp. 77–78.

11. OH, Part I, p. 43 (72).

12. OH, Part I, p. 71 (351).

13. RD and Jean Dubos, *White Plague*, p. 157.

14. RD, "The Initiation of Growth of Certain Facultative Anaerobes as Related to Oxidation-Reduction Processes in the Medium," *JEM* 49 (1929): 559–573; RD, "The Relation of the Bacteriostatic Action of Certain Dyes to Oxidation-Reduction Processes," *JEM* 49 (1929): 575–592; and RD, "The Role of Carbohydrates in Biological Oxidations and Reductions. Experiments with Pneumococcus," *JEM* 50 (1929): 143–160.

15. Harold Ginsberg, [Review of *The Professor, the Institute and DNA*], *ASM News* 43 (1977): 671–672.

16. Alvin F. Coburn to RD, n.d. [1971 letter for an unpublished seventieth birthday festschrift for RD], author's papers.

17. RD, DO 42 (Summer 1973): 378–382.

18. RD to RS, n.d. [Spring 1929], RS Papers.

19. RD to RS, n.d. [Fall 1929], RS Papers.

20. RD to RP, 11 January 1930, RP Letters.

21. Ibid.

22. Louis Pasteur and Jules Joubert, "Charbon et septicémie," *Comptes Rendus des Séances de l'Académie des Sciences* 85 (1877): 101–105.

23. Paul Vuillemin, "Antibiose et symbiose," *Association Française pour l'Avancement des Sciences* Part 2 (1889): 525–543.

24. Selman A. Waksman, *Microbial Antagonisms and Antibiotic Substances*, second edition (New York: The Commonwealth Fund, 1947), p. 331.

In addition, this section relies on other historical discussions of early bacterial antagonisms before the antibiotic era: William Bulloch, *The History of Bacteriology* (London: Oxford University Press, 1938); Howard W. Florey, Ernst Chain, Norman G. Heatley, Margaret A. Jennings, A. Gordon Sanders, Edward P. Abraham, and M. Ethel Florey, *Antibiotics: A Survey of Penicillin, Streptomycin, and Other Antimicrobial Substances from Fungi, Actinomycetes, Bacteria, and Plants* (London: Oxford University Press, 1949); Georges Papacostas and Jean Gaté, *Les associations microbiennes, leurs applications thérapeutiques* (Paris: Librairie Octave Doin, 1928); and Benjamin White, *The Biology of Pneumococcus* (New York: The Commonwealth Fund, 1938).

25. Fleming's Nobel lecture can be found at www.nobel.se.

26. Selman A. Waksman, *The Antibiotic Era* (Tokyo: Waksman Foundation of Japan, 1975), p. 25.

27. Florey et al., *Antibiotics*, pp. 19–26 and 537–554.

28. White, *Biology of Pneumococcus*, pp. 507–521.

29. Harry F. Dowling, *Fighting Infection: Conquests of the Twentieth Century* (Cambridge: Harvard University Press, 1977), p. 106.

30. Gerhard Domagk, "Ein Beitrag zur Chemotherapie der Bakteriellen Infectionen," *Deutsche Medizinische Wochenschrift* 61 (1935): 250–253.

31. Florey et al., *Antibiotics*, p. 422.

32. Selman A. Waksman and Wilburt C. Davison, *Enzymes: Properties, Distribution, Methods, and Applications* (Baltimore: Williams & Wilkins, 1926).

33. L. Rosenthal, "Microbes bacteriolytiques," *Comptes Rendus des Séances de la Société de Biologie* 92 (1925): 78–79; L. Rosenthal and F. Duran-Reynals, "Tyrothrix scaber et flore intestinale," *Comptes Rendus des Séances de la Société de Biologie* 94 (1926): 309; L. Rosenthal and F. Duran-Reynals, "Le sort de Tyrothrix scaber dans l'organisme après son introduction parenterale," *Comptes Rendus des Séances de la Société de Biologie* 94 (1926): 1059; and Jacques Bronfenbrenner, *BSD* 16 (1927–1928): 98–102.

34. Maclyn McCarty, interview with the author, New York City, 5 October 1988.

35. RD to RS, n.d. [Spring 1930], RS Papers.

36. RS to RD, 25 June 1930, RS Papers.

37. Oswald T. Avery and RD, "The Specific Action of a Bacterial Enzyme on Pneumococci of Type III," *Science* 72 (1930): 151–152. The only published excerpt from the OH describes some personal details of this discovery: Saul Benison, "René

Dubos and the Capsular Polysaccharide of Pneumococcus: An Oral History Memoir," *Bulletin of the History of Medicine* 50 (1976): 459–477.

38. RD and Oswald T. Avery, "Decomposition of the Capsular Polysaccharide of Pneumococcus Type III by a Bacterial Enzyme," *JEM* 54 (1931): 51–71; and Oswald T. Avery and RD, "The Protective Action of a Specific Enzyme against Type III Pneumococcus Infection in Mice," *JEM* 54 (1931): 73–89.

39. Kenneth Goodner, RD, and Oswald T. Avery, "The Action of a Specific Enzyme upon the Dermal Infection of Rabbits with Type III Pneumococcus," *JEM* 55 (1932): 393–404.

40. Avery and RD, "Protective Action"; and Martin Burger, *Bacterial Polysaccharides, Their Chemical and Immunological Aspects* (Springfield, Illinois: Charles C. Thomas, 1950), pp. 34–56.

41. RD to RP, 23 January 1933, RP Letters.

42. OH, Part I, p. 57 (86).

43. RD, "Factors Affecting the Yield of Specific Enzyme in Cultures of the Bacillus Decomposing the Capsular Polysaccharide of Type III Pneumococcus," *JEM* 55 (1932): 377–391.

44. OH, Part I, p. 58 (87). Independently in 1930, and unknown to RD, the phenomenon of adaptive enzyme formation was also discovered by Henning Karström in studies of carbohydrate metabolism in Gram-negative enteric bacteria: "Über die Enzymbildung in Bakterien und über Einige Physiologische Eigenschaften der Untersuchten Bakterienarten" (Ph.D. dissertation, Helsingfors, Finland).

45. RD, "Utilization of Selective Microbial Agents in the Study of Biological Problems," *The Harvey Lectures* 35 (1939–1940): 223–242; and RD, "The Adaptive Production of Enzymes by Bacteria," *Bacteriological Reviews* 4 (1940): 1–16.

46. Thomas Francis, Jr. and Edward E. Terrell, "Experimental Type III Pneumococcus Pneumonia in Monkeys. I. Production and Clinical Course," *JEM* 59 (1934): 609–640.

47. RD, *The Professor*, pp. 97–98.

48. Thomas Francis, Jr., Edward E. Terrell, RD, and Oswald T. Avery, "Experimental Type III Pneumococcus Pneumonia in Monkeys. II. Treatment with an Enzyme which Decomposes the Specific Capsular Polysaccharide of Pneumococcus Type III," *JEM* 59 (1934): 641–668.

49. RD, *Celebrations of Life*, p. 184.

50. Papacostas and Gaté, *Les associations microbiennes.*

51. RD, "Domesticating the Microbe," in *New Worlds in Medicine*, edited by Harold Ward (New York: McBride, 1946), pp. 339–345; and Selman A. Waksman,

"Biological Aspects of Antibiotics," in *Frontiers in Medicine* (New York: Columbia University Press, 1951), pp. 99–119.

52. Selman A. Waksman, "Dr. René J. Dubos—A Tribute," *Journal of the American Medical Association* 174 (1960): 111–113.

53. RD to RP, 23 January 1933, RP Letters.

54. Alfred E. Cohn Papers, Record Group 450 C661, Box 15, Folder 13, RUA.

55. RD, *Mirage of Health*, p. 22.

56. Benedict F. Massell, *Rheumatic Fever and Streptococcal Infection: Unraveling the Mysteries of a Dread Disease* (Boston: Harvard University Press for The Francis A. Countway Library of Medicine, 1997).

57. William G. MacCallum, *The Pathology of the Pneumonia in the United States Army Camps during the Winter of 1917–1918* (New York: The Rockefeller Institute for Medical Research Monograph Number 10, 1919).

58. A.R. Dochez, Oswald T. Avery, and Rebecca C. Lancefield, "Studies on the Biology of Streptococcus. I. Antigenic Relationships between Strains of Streptococcus haemolyticus," *JEM* 30 (1919): 179–213. This was Lancefield's initial encounter with the bacterium whose immense diversity and complexity she would master over the next five decades.

59. Homer F. Swift, C.L. Derick, and C.H. Hitchcock, "Rheumatic Fever as a Manifestation of Hypersensitiveness (Allergy or Hyperergy) to Streptococci," *Transactions of the Association of American Physicians* 43 (1928): 192–203.

60. RD to RP, 7 January 1932, RP Letters.

61. Alvin F. Coburn, *The Factor of Infection in the Rheumatic State* (Baltimore: Williams and Wilkins, 1931). A similar report appeared in England about the same time and was based on bacterial throat flora in a small group of rheumatic children: W.R.F. Collis, "Acute Rheumatism and Haemolytic Streptococci," *The Lancet* i (1931): 1341–1345.

62. RD to RS, 7 June 1932, RS Papers.

63. RD to RS, n.d. [Summer 1932], RS Papers.

64. During the 1940s, RD was sufficiently concerned about getting a streptococcal infection that he did not permit experiments with this bacterium, and anyone with a strep throat infection was sent home. Gloria Ross Lewis, a technician of Bernard Davis in the Dubos laboratory, conversation with the author, New York City, 9 October 1993.

65. RD to RS, 20 May 1933, RS Papers.

66. Alfred E. Cohn, and Homer F. Swift, "Electrocardiographic Evidence of Myocardial Involvement in Rheumatic Fever," *JEM* 39 (1924): 1–36.

67. Homer F. Swift, "The Art and Science of Medicine," *Science* 68 (1928): 167–171.

68. F.W. Denny, L.W. Wannamaker, W.R. Brink, C.H. Rammelkamp, Jr., and E.A. Custer, "Prevention of Rheumatic Fever. Treatment of the Preceding Streptococcal Infection," *Journal of the American Medical Association* 143 (1950): 151–153.

69. Maclyn McCarty, interview with the author, New York City, 13 April 2000.

70. RD, *The Professor,* p. 66.

71. Biographical details concerning Marie Louise Bonnet are drawn from transcripts, yearbooks, and correspondence in the archives of Hollins University, Roanoke, Virginia; The Masters School, Dobbs Ferry, New York; the registrar's office of Columbia University, New York; and RD Papers, Box 1, and Box 120, Folder 2. On several occasions and in various ways, Dubos described the tuberculosis that struck Marie Louise, including to the author, in the Oral History and in *Medical World News* (1975), but perhaps most intimately in an interview by Robert Serrou, "Un grand savant français parle/René Dubos," *Paris Match,* 4 April 1980, pp. 3–10.

72. RD to Adéline Dubos, 14 January 1934, RD Family Papers.

73. RD to Adéline Dubos, 1 April 1934, RD Family Papers.

74. RD to Adéline Dubos, n.d. [Spring 1936], RD Family Papers.

75. RD to RS, 26 December 1936, RS Papers.

76. RD, *The Professor,* p. 80.

77. RD to RS, 26 December 1936, RS Papers.

78. RD, "Immunization of Experimental Animals with a Soluble Antigen Extracted from Pneumococci," *JEM* 67 (1938): 799–808.

79. When serum therapy using the S III enzyme was combined with sulfa drugs, the mortality in type III pneumonia patients dropped to 5 percent. Herbert S. Gasser, "Report to the Corporation for 1939–1940," *BSD* 28 (25 October 1940): 18; and Kenneth Goodner, Frank L. Horsfall, Jr., and RD, "Type-Specific Antipneumococcic Rabbit Serum for Therapeutic Purposes. Production, Processing, and Standardization," *Journal of Immunology* 33 (1937): 279–295.

80. Roderick Heffron, *Pneumonia with Special Reference to Pneumococcus Lobar Pneumonia* (New York: The Commonwealth Fund, 1939).

81. Of historic interest, Swift tested the sulfa drugs in rheumatic fever patients and found they had "no demonstrable beneficial influence," some toxicity, and in a few cases magnified rheumatic symptoms. As a result, he warned against their use. Homer F. Swift, *BSD* 26 (1937–1938): 211–221.

82. Oswald T. Avery, *BSD* 15 (1926–1927): 519.

83. Jones suggested the term *tetranuclease* to indicate the ferment exerting its activity on a tetranucleotide: "On the Formation of Guanylic Acid from Yeast Nucleic Acid," *Journal of Biological Chemistry* 12 (1912): 31–35; and RD, "The Decomposition of Yeast Nucleic Acid by a Heat-Resistant Enzyme," *Science* 85 (1937): 549–550.

84. RD and R.H.S. Thompson, "The Decomposition of Yeast Nucleic Acid by a Heat-Resistant Enzyme," *Journal of Biological Chemistry* 124 (1938): 501–510; and idem, "The Isolation of Nucleic Acid and Nucleoprotein Fractions from Pneumococci," *Journal of Biological Chemistry* 125 (1938): 65–74.

85. RD, "Enzymatic Analysis of the Antigenic Structure of Pneumococci," *Ergebnisse der Enzymforschung* 8 (1939): 135–148.

86. RD and Benjamin F. Miller, "The Production of Bacterial Enzymes Capable of Decomposing Creatinine"; idem, "Studies on the Presence of Creatinine in Human Blood"; and idem, "Determination by a Specific, Enzymatic Method of the Creatinine Content of Blood and Urine from Normal and Nephritic Individuals," *Journal of Biological Chemistry* 121 (1937): 429–446, 447–456, and 457–464.

87. Gasser, "Report on the Activities."

88. Merrill W. Chase, "René Jules Dubos, Bacteriologist Extraordinaire," 1988, 42 typescript pages, author's papers.

89. Rollin D. Hotchkiss, "René Dubos: The Early Years,"10 December 1982, RD Papers, Box 1, Folder 7.

90. RD and Jean-Paul Escande, *Quest: Reflections on Medicine, Science, and Humanity*, translated by Patricia Ranum (New York: Harcourt Brace Jovanovich, 1980) p. 62.

91. Rollin D. Hotchkiss, "From Microbes to Medicine: Gramicidin, René Dubos, and the Rockefeller," in *Launching the Antibiotic Era*, edited by Carol L. Moberg and Zanvil A. Cohn (New York: The Rockefeller University Press, 1990), pp. 2–3.

92. For background and perspectives on this recent history of modern antibiotics, the following sources are helpful: Dowling, *Fighting Infections*; Iago Galdston, editor, *The Impact of the Antibiotics on Medicine and Society* (New York: International Universities Press, Inc., 1958); Thomas H. Grainger, *A Guide to the History of Bacteriology* (New York: The Ronald Press Company, 1958); Wallace E. Herrell, *Penicillin and Other Antibiotic Agents* (Philadelphia: W.B. Saunders Company, 1945); Moberg and Cohn, *Launching the Antibiotic Era*; John Parascandola, editor, *The History of Antibiotics: A Symposium* (Madison, Wisconsin: American Institute of the History of Pharmacy, 1980); Kenneth B. Raper, "A Decade of Antibiotics in America," *Mycologia* 44 (1952): 1–59; Morris J. Vogel and Charles E. Rosenberg, editors, *The Therapeutic Revolution: Essays in the Social History of American Medicine* (Philadelphia: University of Pennsylvania Press, Inc., 1979); Waksman, *Microbial Antagonisms*;

Waksman, *The Antibiotic Era*; and Henry Welch and Charles N. Lewis, *Antibiotic Therapy* (New York: Medical Encyclopedia, Inc., 1953).

93. RD, "Bactericidal Effect of an Extract of a Soil Bacillus on Gram Positive Cocci," *Proceedings of the Society for Experimental Biology and Medicine* 40 (1939): 311–312; RD, "Studies on a Bactericidal Agent Extracted from a Soil Bacillus. I. Preparation of the Agent. Its Activity *in vitro*," and "II. Protective Effect of the Bactericidal Agent against Experimental Pneumococcus Infections in Mice," *JEM* 70 (1939): 1–17; and RD and Carlo Cattaneo, "Studies on a Bactericidal Agent Extracted from a Soil Bacillus. III. Preparation and Activity of a Protein-Free Fraction," *JEM* 70 (1939): 249–256.

Even as a bacteriologist, RD did not undertake experiments to identify the bacterial species from which either the 1930 or the 1939 antibacterial agents were derived. In the case of the S III bacillus, the identification was made in 1934 at the New York State Department of Health. Grace M. Sickles and Myrtle Shaw proposed the name *Bacillus palustris* (from the Latin word meaning marsh or swamp) for what was deemed an unusual and previously unknown organism: "A Systematic Study of Microorganisms which Decompose the Specific Carbohydrates of the Pneumococcus," *Journal of Bacteriology* 28 (1934): 415–431.

RD credited identification of the 1939 microbe as *Bacillus brevis* (strain B.G.) to Nathan Smith, Ruth Gordon, and Francis Clark at the U.S. Department of Agriculture, which in the mid-1930s initiated a systematic collection of the genus *Bacillus*. RD and Rollin D. Hotchkiss, "The Production of Bactericidal Substances by Aerobic Sporulating Bacilli," *JEM* 73 (1941): 629–640.

Today the American Type Culture Collection has renamed *B. brevis* as *Brevibacillus parabrevis* (ATCC Number 8185; www.atcc.org). Ruth E. Gordon, William C. Haynes, and C. Hor-Nay Pang, *The Genus* Bacillus (Washington, D.C.: U.S. Department of Agriculture, Agricultural Research Service, Handbook Number 427, 1973).

As part of his research on *B. brevis*, RD obtained and compared his strain with six *Tyrothrix* cultures from the Lister Institute in England that had been isolated by Duclaux in 1887, but he found these cultures were much less bactericidal than his original *B. brevis*; the ATCC has now identified these *Tyrothrix* cultures as strains of *Bacillus subtilis*. Of historic note, enzymes from *B. subtilis* added to laundry detergents in the 1960s were a subject of RD's research and warnings against their use (chapter 5).

94. Hotchkiss, "From Microbes to Medicine."

95. Rollin D. Hotchkiss and RD, "Bactericidal Fractions from an Aerobic Sporulating Bacillus," *Journal of Biological Chemistry* 136 (1940): 803–804.

96. Rollin D. Hotchkiss and RD, "The Isolation of Bactericidal Substances from Cultures of Bacillus brevis," *Journal of Biological Chemistry* 141 (1941): 155–162.

97. Dubos and Hotchkiss, "Production of Bactericidal Substances"; and idem, "Origin, Nature and Properties of Gramicidin and Tyrocidine," *Transactions and Studies of the College of Physicians of Philadelphia* 10 (1942): 11–19.

98. Ralph B. Little, RD, and Rollin D. Hotchkiss, "Gramicidin, Novoxil, and Acriflavine for the Treatment of the Chronic Form of Streptococcic Mastitis," *Journal of the American Veterinary Medical Association* 98 (1941): 189–199; and Ralph B. Little, RD, Rollin D. Hotchkiss, C.W. Bean, and W.T. Miller, "The Use of Gramicidin and Other Agents for the Elimination of the Chronic Form of Bovine Mastitis," *American Journal of Veterinary Research* 2 (1941): 305–312.

99. French Nobel laureate François Jacob successfully tested tyrothricin against tuberculosis. For his M.D. thesis in 1947 at the Faculté de Paris, he produced tyrothricin from strains of bacteria sent by Dubos and tested its clinical applications in tuberculosis and several other diseases. Jacob concluded that tyrothricin's drawbacks (toxicity to red blood cells and insolubility in water) could be "limited or overcome" and that "tyrothricin was unjustly eclipsed by penicillin." P.J.F. Broch, François Jacob, R. Courtade, and P. Bocquet, *Un proche parent de la pénicilline. La tyrothricine, fabrication, applications cliniques, industrielles et agricoles* (Paris: Vigot Frères, 1947); P.J.F. Broch and François Jacob, "Quelques aspects particuliers de la sécretion antibiotique du Bacillus brevis," *Fourth International Congress for Microbiology (1947): Report of Proceedings*, edited by Mogens Bjorneboe (Copenhagen: Rosenkilde and Bagger, 1949), pp. 139–140; and François Jacob, *The Statue Within: An Autobiography*, translated by Franklin Philip (New York: Basic Books, Inc., 1988), pp. 192–193.

100. Oswald T. Avery, *BSD* 27 (1938–1939): 151–174; and Homer F. Swift, *BSD* 27 (1938–1939): 220–233.

101. George K. Hirst, "The Effect of a Polysaccharide-Splitting Enzyme on Streptococcal Infection," *JEM* 73 (1941): 493–506.

102. Colin M. MacLeod, George S. Mirick, and Edward C. Curnen, Jr., "Toxicity for Dogs of a Bactericidal Substance Derived from a Soil Bacillus," *Proceedings of the Society for Experimental Biology and Medicine* 43 (1940): 461–463,

103. RD, *BSD* 29 (1940–1941): 150; and *BSD* 30 (1941–1942): 158. This unpublished research is mentioned in a later publication: RD, "Tyrothricin, Gramicidin, and Tyrocidine—Twenty Years Later, With Remarks on the Tissue Factors Which Affect Chemotherapy," *Antibiotics Annual 1959–60* (NY: Antibiotica, Inc., 1960), pp. 343–349.

104. Gasser, "Report to the Corporation for 1939–1940."

105. Hotchkiss gave slight variants of this story in "René Dubos: The Early Years," and in "From Microbes to Medicine."

On 12 August 1943, the Institute abandoned this patent application, stating, "It would seem that the original purposes of the application which was filed on January 5, 1940 had been served, and that, under the circumstances there does not appear to be any advantage to be gained in pressing the claims for patent." Record Group 210.3, Box 12, Folder "Gramicidin 1939–1943, 1949," RUA.

106. Wallace E. Herrell and Dorothy Heilman, "Experimental and Clinical Studies on Gramicidin," *Journal of Clinical Investigation* 20 (1941): 433; and Charles H. Rammelkamp and Chester S. Keefer, "Observations on the Use of 'Gramicidin' (Dubos) in the Treatment of Streptococcal and Staphylococcal Infections," *Journal of Clinical Investigation* 20 (1941): 433–434. A few days after this annual meeting Herrell and Heilman submitted their fuller report "Experimental and Clinical Studies on Gramicidin," *Journal of Clinical Investigation* 20 (1941): 583–591.

107. A full record of this early clinical research appears in Herrell, *Penicillin and Other Antibiotic Agents*, pp. 267–302.

108. Charles H. Rammelkamp, "Use of Tyrothricin in the Treatment of Infections; Clinical Studies," *War Medicine* 2 (1942): 830–846; and Chester S. Keefer, RD, John S. Lockwood, and E. Kennerly Marshall, Jr., "Symposium on War Medicine. Chemotherapy. I. Pharmacology and Toxicology," *Clinics* 2 (1944): 1077–1093.

109. Charles H. Rammelkamp, "Observations on the Resistance of Staphylococcus aureus to the Action of Tyrothricin," *Proceedings of the Society for Experimental Biology and Medicine* 49 (1942): 346–350.

110. R. Lloyd Phillips and Lucy H. Barnes, "Development of Resistance in Staphylococci to Natural Inhibitory Substances," *Journal of The Franklin Institute* 233 (1942): 396–401; S.J. Crowe, A.M. Fischer, A.T. Ward, Jr., and M.K. Foley, "Penicillin and Tyrothricin in Otolaryngology Based on a Bacteriological and Clinical Study of 118 Patients," *Annals of Otology, Rhinology & Laryngology* 52 (1943): 541–572; and Rollin D. Hotchkiss, "Gramicidin, Tyrocidine, and Tyrothricin," *Advances in Enzymology* 4 (1944): 153–199.

111. B.A. Wallace, "Gramicidin Channels and Pores," *Annual Review of Biophysics and Biophysical Chemistry* 19 (1990): 127–157.

112. Corner recounted that, early in the Institute's first decade, Flexner wanted to publicize its scientific advances. Chairman Frederick Gates thought this unwise and admonished Flexner, saying that he wanted "to hold the Institute strictly to its business and to stop the ears of everybody about it from hearing any voice but the voice of science." Showing the importance and humaneness of the research was necessary, however, when dealing with activists, such as the antivivisectionists. Nonetheless, minimizing publicity was important to the scientific directors since successes in basic research were not easy to explain and hints of substantial benefits to public health could be greatly misinterpreted. While experienced science reporters like Walter Sullivan of *The New York Times* may have complained about the Institute's anti-publicity policy, they were able to glean good news stories and occasional interviews by attending major scientific congresses. Corner, *A History of The Rockefeller Institute*, pp. 158–160.

113. Fiorello H. LaGuardia, "Official Opening of the Congress," in *Third International Congress for Microbiology: Report of Proceedings*, edited by M. Henry Dawson (New York: International Association of Microbiologists, 1940), pp. 20–27.

114. "Pneumonia Yields to New Chemical," *The New York Times*, 9 September 1939; and "Sugar Coated Germs," *The New York Times*, 17 September 1939.

115. Some critical sources relevant to the history of penicillin include Lennard Bickel, *Rise Up to Life: A Biography of Howard Walter Florey Who Gave Penicillin to the World* (London: Angus and Robertson, 1972); Ronald W. Clark, *The Life of Ernst Chain: Penicillin and Beyond* (New York: St. Martin's Press, 1985); Ronald Hare, *The Birth of Penicillin and the Disarming of Microbes* (London: Allen and Unwin, 1970); Eric Lax, *The Mold in Dr. Florey's Coat* (New York: Henry Holt, 2004); Gwyn Macfarlane, *Alexander Fleming: The Man and the Myth* (Cambridge: Harvard University Press, 1984); idem, *Howard Florey: The Making of a Great Scientist* (Oxford: Oxford University Press, 1979); John C. Sheehan, *The Enchanted Ring: The Untold Story of Penicillin* (Cambridge: M.I.T. Press, 1982); and Trevor I. Williams, *Howard Florey: Pencillin and After* (Oxford: Oxford University Press, 1984).

116. Milton Wainwright and Harold T. Swan, "C.B. Paine and the Earliest Surviving Clinical Records of Penicillin Therapy," *Medical History* 30 (1986): 42–56.

117. OH, Part I, p. 86 (452–453).

118. According to Macfarlane, biographer of both Fleming and Florey, Fleming attended the 1939 Congress in New York. Florey, however, traveled to the United States only twice, in 1925–1926 and again in 1941. The *Proceedings* of the Congress lists Fleming's participation but does not contain Florey's name either on the program or as attending the Congress. In addition, according to Norman Heatley, who was the youngest member of the penicillin team, the entry in his personal diary notes that he met with Professor Florey at the Dunn School on 4 September 1939. Heatley, letter to the author, November 1989.

119. Quoted in Macfarlane, *Florey,* p. 299.

120. Quoted in Clark, *Chain,* p. 34.

121. Quoted in Macfarlane, *Florey*, p. 299.

122. Florey et al., *Antibiotics*, p. 637.

123. Williams, *Florey,* p. 95.

124. Quoted in Florey et al., *Antibiotics*, p. 637.

125. Ernst Chain, Howard W. Florey, A.D. Gardner, Norman G. Heatley, Margaret A. Jennings, Jean Orr-Ewing, and A. Gordon Sanders, "Penicillin as a Chemotherapeutic Agent," *The Lancet* ii (1940): 226–228; and Edward P. Abraham, Ernst Chain, Charles M. Fletcher, A.D. Gardner, Norman G. Heatley, Margaret A. Jennings, and Howard W. Florey, "Further Observations on Penicillin," *The Lancet* ii (1941): 177–188.

In the summer of 1941, when Florey and Heatley visited the United States to secure help in producing penicillin on a large scale, it was Dubos who informed Florey of the Mayo Clinic's interest in studies of antibiotic agents, according to Herrell, "including our preliminary studies on penicillin. . . . I was glad to receive word from Dubos saying that Florey was coming to visit our laboratories after his stay in Peoria. . . . Florey expressed satisfaction with the studies we had under way," and as a result had Heatley supply Herrell with one hundred milligrams of penicillin, half of which was prepared in Oxford and the other half recovered from the urine of an Oxford patient. Herrell, *Penicillin*, p. 7.

126. Virginia Cameron and Esmond R. Long, *Tuberculosis Medical Research: National Tuberculosis Association, 1904–1955* (New York: National Tuberculosis Association, 1959), p. 38.

127. Waksman, *The Antibiotic Era*, p. 11.

128. Cameron and Long, *Tuberculosis Medical Research*, p. 39.

129. Selman A. Waksman, "Fred Beaudette as Friend and Scientist,"in *Perspectives in Virology*, edited by Morris Pollard (New York: John Wiley and Sons, 1958), pp. 1–6.

130. Albert Schatz, Elizabeth Bugie, and Selman A. Waksman, "Streptomycin, a Substance Exhibiting Antibiotic Activity against Gram-positive and Gram-negative Bacteria," *Proceedings of the Society for Experimental Biology and Medicine* 55 (1944): 66–69.

131. Waksman, *My Life with Microbes*, p. 226.

132. This prize was also shared by Vincent du Vigneaud, who discovered the hormone oxytocin.

133. The Nobel Archives in Sweden has recently released information on nominations made during its first fifty years. According to their records, RD was nominated four times before 1952 for the Prize in Physiology and Medicine, in 1944, 1945, 1946, and 1949.

134. The Nobel Foundation and W. Odelberg, editors, *Nobel, the Man and His Prizes*, third edition (New York: American Elsevier, 1972) p. 201. More recently, Peter Reichard, who was close to members of the Nobel Committee, called the Avery work "the greatest biological discovery of the century." Peter Reichard, "Osvald [*sic*] T. Avery and the Nobel Prize in Medicine," *Journal of Biological Chemistry* 277 (2002): 13355–13362.

135. Thomas S. Kuhn, *The Structure of Scientific Revolutions* (Chicago: University of Chicago Press, 1962), p. 134.

136. George Mackaness to RD, 4 December 1981, RD Papers, Box 128, Folder 4.

137. The following general interest publications, in chronological order, describe the Dubos discoveries: H. Dyson Carter, "Miracles from Mud," *Maclean's Magazine* 54 (15 August 1941): 13, 31–32; John Pfeiffer, "Germ-Killers from the Earth: The Story of Gramicidin," *Harper's* 184 (March 1942): 431–437; John D. Ratcliff, *Yellow Magic* (New York: Random House, 1945), pp. 30–39; [Archibald MacLeish,

according to Jean Dubos], "Medicine from Earth," *Fortune* 34 (July 1946): 98–105, 208, 210; Samuel Epstein and Beryl Williams Epstein, *Miracles from Microbes; The Road to Streptomycin* (New Brunswick, New Jersey: Rutgers University Press, 1946), pp. 80–98; I.B. Cohen, "Conditions of Scientific Discovery," *Science, Servant of Man* (Boston: Little, Brown, 1948), pp. 16–35; and George W. Gray, "The Antibiotics," *Scientific American* 181 (August 1949): 26–35.

138. Tom Rivers, *Reflections on a Life in Medicine and Science: An Oral History Memoir Prepared by Saul Benison* (Cambridge: M.I.T. Press, 1967), p. 199.

139. Robert Taylor, *Saranac: America's Magic Mountain* (Boston: Houghton, Mifflin, 1986), pp. 271–273.

140. Kenneth W. Wright, James Monroe, and Frederick Beck, "A History of the Ray Brook State Tuberculosis Hospital," *New York State Journal of Medicine* 90 (1990): 406–413.

141. Marie Louise Dubos to RD, 1 July 1940, RD Papers, Box 128.

142. Madeleine Vorwald to RD, 17 November 1976, RD Papers.

143. Marie Louise Dubos to RD, 25 March 1942, RD Papers, Box 128.

144. In the OH, RD related that Marie Louise's tuberculosis "was too advanced to be cured," but she had "improved sufficiently so that she in time came to lead a semi-normal life. When that stage was reached, it was decided she might come back to New York. . . . However, within about two hours after reaching New York City she went into a state of panic that she was no longer able to cope with what was expected of a normal life . . . within one week she had another very acute attack of her disease and perhaps a week or two later, she died of overwhelming tuberculosis." OH, Part II, pp. 106 (559–560).

In contrast, when mail service resumed to France in 1944, Dubos wrote his mother that Marie Louise came home to New York in good health (*bonne santé)* but died a week after her return. (RD Family Papers).

Trying to understand her cause of death is important because of the research path RD took when he devoted his laboratory to tuberculosis. If she died from overwhelming or miliary tuberculosis, this fact should be considered an infectious cause of death, although the death certificate may have not recorded it. However, another noninfectious disease may have been the cause, in particular, one related to her rheumatic heart disease, such as congestive heart failure, bacterial endocarditis, or a rheumatic pneumonitis.

Still another possible cause of death is a rare concomitant infection of tuberculosis with other microbes, such as the fungus *Aspergillus*. This ubiquitous and normally benign fungus can cause a noninfectious systemic disease and death by invading bronchi damaged by tuberculosis. Wright reports in his history of Ray Brook that the sanatorium also cared for patients with various fungal diseases of the lungs, among them aspergillosis. Curiously, this disease is one that Dubos studied with the Ray

Brook sanatorium scientists and Keith Porter after her death, even seeking an antibiotic from its causal organism. For reasons that are not clear, he also cited this disease as an example of the constellation of metabolic events that turn an infection into overt disease. In retrospect, the phenomena of latent and concomitant infections, which became major research interests in his laboratory, may have been spurred more by such multiple complications that beset Marie Louise at the end of her life.

145. RD, "Le passé vivant," 28 April 1943, eight handwritten pages, RD Papers, Box 120, Folder 2.

146. RD to Oswald T. Avery, 18 July 1942, Personal Papers of Maclyn McCarty.

147. George Corner, in *A History of The Rockefeller Institute*, describes some of the research by scientists who took part in the Institute's national defense effort during World War II, pp. 518–532.

148. A substantial body of secondary literature exists on this topic. Proving that scientists did not foster germ warfare, when claims can be fabricated that they did, is probably an impossible task. The following historical sources, alphabetical by author, seem objective about the wartime program: E.C. Andrus, D.W. Bronk, G.A. Carden, Jr., C.S. Keefer, J.S. Lockwood, J.T. Wearn, and M.C. Winternitz, *Advances in Military Medicine, Made by American Investigators Working Under the Sponsorship of the Committee on Medical Research*, two volumes (Boston: Little, Brown and Company, 1948); Barton J. Bernstein, "America's Biological Warfare Program in the Second World War," *The Journal of Strategic Studies* 11 (1988): 292–313; George W. Christopher, Theodore J. Cieslak, Julie A. Pavlin, Edward M. Eltzen, Jr., "Biological Warfare: A Historical Perspective," *Journal of the American Medical Association* 278 (1997): 412–418; Rexmond D. Cochrane, *The National Academy of Sciences: The First Hundred Years, 1863–1963* (Washington, D.C.: National Academy of Sciences, 1978); Norman M. Covert, *Cutting Edge: A History of Fort Detrick, Maryland, 1943–1993* (Fort Detrick, Maryland: The Headquarters, 1993); Stockholm International Peace Research Institute, *The Problem of Chemical and Biological Warfare*, six volumes (Stockholm: Almqvist & Wiskell, 1971–1975).

149. Cochrane, *The National Academy of Sciences*, p. 413; "Historical Report, November 1944–Final," a declassified top secret typescript (NAS Archives: Committees on Biological Warfare Records Group, Box 5, Series 4, War Research Service); E.B. Fred, "Memorandum for Mr. Merck," 19 January 1943 (NAS Archives: Committees on Biological Warfare, Box 4, War Research Service: Beginning of Program, 1942–1943, Series 4, War Research Service).

150. OH, Part I, p. 99 (555).

151. A complete list of OSRD contracts, investigators, and their institutions appears in Andrus et al., *Advances in Military Medicine*, volume two, pp. 831–882. Of the seven contracts for bacillary dysentery, four were for producing a vaccine: two at The Rockefeller Institute (to RD and to Walther Goebel), one at the University of

Pennsylvania, and another at Children's Hospital in Cincinnati. The three other contracts were to develop drugs to control bacillary dysentery and they were assigned to the University of Chicago, the University of Southern California, and the National Institutes of Health.

152. For the dysentery project, RD was assigned two OSRD contract numbers. At Harvard, he and Henry P. Treffers were investigators on contract OEMcmr-170. The contract was continued as OEMcmr-449 when RD returned to Rockefeller in 1944, and it ended in 1945; it was for the preparation of a vaccine. A different contract, OEMcmr-499, was issued to Treffers when he moved to Yale University and was for the preparation of a synthetic antigen against shigella.

153. RD, Henry D. Hoberman, and Cynthia Pierce, "Some Factors Affecting the Toxicity of Cultures of Shigella dysenteriae," *Proceedings of the National Academy of Sciences U.S.A.* 28 (1942): 453–458.

154. RD to E. Cowles Andrus, 20 July 1943, OSRD E-165 "General Records, Dubos, René."

155. "Historical Report, November 1944–Final," Section II, pp. 21–22.

156. RD to E. Cowles Andrus, 19 January 1943; and James S. Simmons to E. Cowles Andrus, 22 January 1943, OSRD E-165 "General Records, Dubos, René."

157. RD, June Hookey Straus, and Cynthia Pierce, "The Multiplication of Bacteriophage in vivo and Its Protective Effect Against an Experimental Infection with Shigella dysenteriae," *JEM* 78 (1943): 161–168.

158. Felix d'Herelle, *The Bacteriophage and its Clinical Applications*, translated by George H. Smith (Springfield, Illinois: Charles C. Thomas, 1930).

159. Thomas D. Brock pointed out in 1961 that "all trials of bacteriophage use as therapeutants have been unsuccessful": *Milestones in Microbiology* (Englewood Cliffs, N.J.: Prentice-Hall, 1961), p. 159. More recently, attempts are being made to use phage therapy to treat infections that are refractory to antibiotics. One such report describes extensive controlled use of bacteriophage for prophylaxis and treatment of *E. coli* infections in farm animals: H. William Smith, M.B. Huggins, and K.M. Shaw, "The Control of Experimental Escherichia coli Diarrhoea in Calves by Means of Bacteriophage," *Journal of General Microbiology* 133 (1987): 1111–1126. Two other reports from The Rockefeller University laboratory of Vincent Fischetti are reminiscent of a Dubosian search and show that enzymes produced by bacteriophages kill only the targeted disease bacteria on contact: D. Nelson, L. Loomis, and V.A. Fischetti, "Prevention and Elimination of Upper Respiratory Colonization of Mice by Group A Streptococci by Using a Bacteriophage Lytic Enzyme," *Proceedings of the National Academy of Sciences U.S.A.* 98 (2001): 4107–4112; and R. Schuch, D. Nelson, and V.A. Fischetti, "A Bacteriolytic Agent that Detects and Kills Bacillus anthracis," *Nature* 418 (2002): 884–889.

160. RD to A.N. Richards, 10 February 1944, OSRD E-165 "General Records, Dubos, René."

161. A memorandum from George Merck, co-director of the War Research Service (WRS), to the chief of the Chemical Warfare Service (CWS), dated 9 June 1944, recommended the CWS assume responsibility for dysentery among other War Research Service projects. "Historical Report, November 1944–Final," Section III, p. 73.

Despite this recommendation, no records have been found to indicate that the dysentery organism was developed as a biological weapon after the WRS ended in 1944 or that RD worked on a CWS contract. This has been corroborated by personal communication of the author with Riley Housewright, scientific director at Camp Detrick beginning in 1943; William Patrick, development director of weapons at Camp Detrick beginning in 1950; Colonel Arthur Anderson at Fort Detrick, who pursued archival searches on this question in the papers of Norman Covert, historian of Camp Detrick, and the CWS papers at Dugway Proving Ground; and Wilbert Mahoney, archivist of CWS military records in the National Archives. See also Leo P. Brophy, Wyndham D. Miles, and Rexmond C. Cochrane, *The Chemical Warfare Service: From Laboratory to Field* (Washington, D.C.: Office of the Chief of Military History, Department of the Army, 1959).

162. James E. McCormack to RD, 14 December 1944, OSRD E-165 "General Records, Dubos, René."

163. Colin MacLeod, Chief of the CMR's Division of Medicine, offered the Squibb preparations of the dysentery vaccine to the National Institutes of Health "with the possibility that testing of its antigenicity might be made in man sometime in the future." MacLeod to Dr. M.V. Veldee, 27 February 1946, OSRD E-163 "Contract Records, Rockefeller Institute, Dubos, OEMcmr-449."

164. RD and James W. Geiger, "Preparation and Properties of Shiga Toxin and Toxoid," *JEM* 84 (1946): 143–156. After the war projects ended, bacterial toxins became a major area of research in infectious diseases. Walther Goebel in Avery's laboratory studied ways to attenuate the toxicity of the *Shigella* endotoxin. William van Heyningen at Oxford University used RD's methods to obtain a purer and more concentrated exotoxin for vaccine use, but by that time, RD was interested in the role of toxins in converting infection into disease.

165. OH, Part I, pp. 98, 102 (554, 558).

Chapter 3 Tuberculosis and Dilemmas of Modern Medicine

1. OH, part I, p. 118 (574).

2. OH, part I, pp. 114, 118 (570, 574).

3. Tom Rivers, *Reflections on a Life in Medicine and Science: An Oral History Memoir Prepared by Saul Benison* (Cambridge: M.I.T. Press, 1967), p. 399.

4. RD to Thomas F. Rivers, 29 July 1943, Thomas Rivers Papers, American Philosophical Society, Philadelphia, Pennsylvania.

5. *Minutes of the Board of Scientific Directors and of the Executive Committee, The Rockefeller Institute for Medical Research*, 30 October 1943 and 14 January 1944, Record Group 110.2, RUA.

6. Charles Sidney Burwell, "Memorandum for Discussion with René Dubos," 13 October 1943, Dean's Subject Files, 1899–1953, Countway Medical Library, Harvard University; and RD to James B. Conant, President of Harvard University, 1 November 1943, Conant Papers, Harvard University Archives, quoted in Jean A. Curran, *Founders of the Harvard School of Public Health* (New York: Josiah Macy, Jr. Foundation, 1970), pp. 258–259.

7. A. Baird Hastings, *Crossing Boundaries: Biological, Disciplinary, Human. A Biochemist Pioneers for Medicine* (Grand Rapids, Michigan: The Four Corners Press, 1989), pp. 311–312. For a full text of the parable, see RD to A. Baird Hastings, November 1945, RD Papers.

8. Arnold R. Rich, *The Pathogenesis of Tuberculosis* (Springfield, Illinois: Charles C. Thomas, 1944), p. 895.

9. RD, "The Living Thought of Louis Pasteur," *MD* 7 (1952): 381–384, 402.

10. RD, "Microbiology," *Annual Review of Biochemistry* 11 (1942): 659–678.

11. RD, "The Variability of Bacteria," in *The Bacterial Cell*, pp. 135–187; and Hans Zinsser and Stanhope Bayne-Jones, "Variability of Bacteria," in *A Textbook of Bacteriology*, eighth edition (New York: D. Appleton-Century Company, 1939), pp. 129–141.

12. Paul H. de Kruif, "Dissociation of Microbic Species. I. Coexistence of Individuals of Different Degrees of Virulence in Cultures of the Bacillus of Rabbit Septicemia," *JEM* 33 (1921): 773–789; Theobald Smith, "Koch's Views on the Stability of Species among Bacteria," *Annals of Medical History* 4 (1932): 524–530; Sergei Winogradsky, "The Doctrine of Pleomorphism in Bacteriology," *Soil Science* 43 (1937): 327–340; RD, "Bacterial Variability," in *The Professor*, pp. 125–138; RD, "One Century of Bacterial Polymorphism," in *The Role of Culture Collections in the Era of Molecular Biology*, edited by Rita R. Colwell (Washington, D.C.: American Society for Microbiology, 1976), pp. 35–42; and M. Penn and M. Dworkin, "Robert Koch and Two Visions of Microbiology," *Bacteriological Reviews* 40 (1976): 276–283.

13. RD, *The Bacterial Cell*, pp. 135–136.

14. Carl F. Robinow, "Nuclear Apparatus and Cell Structure of Rod-Shaped Bacteria," in RD, *The Bacterial Cell*, pp. 353–377.

15. George A. Jacoby and Gordon L. Archer, "New Mechanisms of Bacterial Resistance to Antimicrobial Agents," *New England Journal of Medicine* 324 (1991):

601–612; Julian Davies, "Inactivation of Antibiotics and the Dissemination of Resistance Genes," *Science* 264 (1994): 375–382.

16. Carol L. Moberg, "René Dubos: A Harbinger of Microbial Resistance to Antibiotics," *Microbial Drug Resistance* 2 (1996): 287–297.

17. Colin M. MacLeod and Giuseppe Daddi, "A 'Sulfapyridine-Fast' Strain of Pneumococcus Type I," *Proceedings of the Society for Experimental Biology and Medicine* 41 (1939): 69–71; Colin M. MacLeod, "Metabolism of 'Sulfapyridine-Fast' and Parent Strains of Pneumococcus Type I," *Proceedings of the Society for Experimental Biology and Medicine* 41 (1939): 215–218; Colin M. MacLeod and George S. Mirick, "Bacteriological Diagnosis of Pneumonia in Relation to Chemotherapy," *American Journal of Public Health* 31 (1941): 34–38; and Frank Horsfall, "Effect of Sulfonamides on Virulence of Pneumococci," *Journal of Clinical Investigation* 21 (1942): 647.

18. MacLeod and Mirick, "Bacteriological Diagnosis," p. 36.

19. RD, "Trends in the Study and Control of Infectious Diseases," *Proceedings of the American Philosophical Society* 88 (1944): 208–213.

20. Joshua Lederberg, Introduction to *Launching the Antibiotic Era: Personal Accounts of the Discovery and Use of the First Antibiotics*, edited by Carol L. Moberg and Zanvil A. Cohn (New York: The Rockefeller University Press, 1990), pp. ix–x.

21. RD, "Mechanism of Drug Fastness," in *The Bacterial Cell*, pp. 318–329.

22. Quoted in English translation by RD, *The Bacterial Cell*, p. 348, from Claude Bernard, *Leçons sur les phénomènes de la vie communs aux animaux et aux végétaux*, facsimile of original 1878 edition (Paris: J. Vrin, 1966), p. 50.

23. Lederberg, Introduction.

24. The symposium "Heredity and Variation in Microorganisms" in July 1946 at Cold Spring Harbor marks something of a watershed between bacterial physiologists and bacterial geneticists. Some luminary scientists in attendance included Thomas Anderson, Oswald Avery, Max Delbrück, A.D. Hershey, Rollin Hotchkiss, Joshua Lederberg, Salvador Luria, André Lwoff, Maclyn McCarty, Barbara McClintock, Ernst Mayr, Jacques Monod, E.L. Tatum, and C.B. van Niel. While the symposium burgeoned with papers on genetic topics, Dubos' presentation dealt with the concept of plasticity presented a year earlier in *The Bacterial Cell*; he was an outsider in suggesting the time had come for doing something about host-parasite interactions. RD, "Variations in Antigenic Properties of Bacteria," *Cold Spring Harbor Symposia on Quantitative Biology* 11 (1946): 60–66.

25. Bernard D. Davis, "Two Perspectives: on René Dubos and on Antibiotic Actions," in Moberg and Cohn, *Launching the Antibiotic Era*, pp. 69–70.

26. Frank Fenner to RD, n.d. [1971 letter for an unpublished seventieth birthday festschrift for RD], author's papers.

27. The scientific descriptions of the tuberculosis research between 1944 and 1960 draw on three main sources: nearly one-hundred laboratory publications; RD's annual reports to the Board of Scientific Directors (*BSD*); and multiple interviews with Merrill Chase, Cynthia Pierce-Chase, Bernard Davis, Henriette Noufflard, Robert Fauve, Ian Maclean Smith, James G. Hirsch, and Zanvil A. Cohn. In addition, these books add perspectives about medical, pathological, bacteriological, and clinical aspects of tuberculosis during the 1940s: F.M. Burnet, *Biological Aspects of Infectious Disease* (Cambridge: University Press, 1940); Virginia Cameron and Esmond R. Long, *Tuberculosis Medical Research: National Tuberculosis Association (1904–1955)* (New York: National Tuberculosis Association, 1959); Georges Canetti, *Le bacille de Koch dans les lésions tuberculeuses du poumon de l'homme* (Paris: Éditions Médicales Flammarion, 1946); Benjamin Goldberg, editor, *Clinical Tuberculosis*, fifth edition (Philadelphia: F.A. Davis Co., 1946); Max B. Lurie, *Resistance to Tuberculosis* (Cambridge: Harvard University Press, 1964); and Rich, *The Pathogenesis of Tuberculosis.*

28. Davis, "Two Perspectives," p. 73.

29. Merrill W. Chase, "René Jules Dubos, Bacteriologist Extraordinaire," 1988, 42 typescript pages, author's papers.

30. Theobald Smith, *Research Into the Causes and Antecedents of Disease, Its Importance to Society* (Boston: The Old Corner Book Store, Inc., 1905).

31. Theobald Smith, *Parasitism and Disease* (Princeton: Princeton University Press, 1934).

32. T. Mitchell Prudden, *The Story of the Bacteria and Their Relations to Health and Disease* (New York: G.P. Putnam's Sons, 1890).

33. "The Award of the Francis B. Trudeau Medal for 1951," *ART* 64 (1951): 323–326.

34. E.L. Opie and J.D. Aronson, "Tubercle Bacilli in Latent Tuberculous Lesions and in Lung Tissues without Tuberculous Lesions," *Archives of Pathology* 4 (1927): 1–21; and E.L. Opie, E.W. Flahiff, and H.H. Smith, "Protective Inoculation Against Human Tuberculosis with Heat Killed Tubercle Bacilli," *American Journal of Hygiene, Section B* 29 (1939): 155–164.

35. E. Cowles Andrus, "Conference with Dr. MacLeod, Dr. Dochez, and Dr. Dubos concerning the Possible Sponsorship by CMR of Research on Tuberculosis," 24 October 1944; E. Cowles Andrus to RD, 19 December 1944, and W. Mansfield Clark to E. Cowles Andrus, 15 March 1945; all are in OSRD E-165 "General Records, Dubos, René."

36. Diran Yegian to Keith R. Porter, 18 April 1945 and 16 July 1945, Keith R. Porter Papers, Box CM 37, Archives, University of Colorado at Boulder Libraries. An unpublished survey of chemical compounds appears in RD, *BSD* 35 (1946–1947): 49–62. The resulting publications on antimycobacterial substances include: W.E.

Doering, RD, D.S. Noyce, and R. Dreyfus, "Metabolic Products of *Aspergillus ustus*," *Journal of the American Chemical Society* 68 (1946): 725; and George Hogeboom and Lyman C. Craig, "Identification by Distribution Studies. VI. Isolation of Antibiotic Principles from *Aspergillus ustus*," *Journal of Biological Chemistry* 162 (1946): 363–368; RD, "Inhibition of Bacterial Growth by Auxins," *Proceedings of the Society for Experimental Biology and Medicine* 63 (1946): 317–319; Alfred Marshak, G.T. Barry, and Lyman C. Craig, "Antibiotic Compound Isolated from the Lichen *Ramalina reticulata*," *Science* 106 (1947): 394–395; Cutting B. Favour, "Lytic Effect of Bacterial Products on Lymphocytes of Tuberculous Animals," *Proceedings of the Society for Experimental Biology and Medicine* 65 (1947): 269–272; and Richard Gruber, RD, Cynthia Pierce, and Harry E. Ungerleider, "4-Caproylamino Diphenylsulfone, 4′-Aminomethylsulfonic Acid Sodium Salt. Pharmacology and Effect in Experimental Tuberculosis," *Journal of Clinical Investigation* 27 (1948): 538.

37. James G. Hirsch and RD, "The Effect of Spermine on Tubercle Bacilli," *JEM* 95 (1952): 191–208; idem, "Chemical Studies on a Basic Peptide Preparation Derived from Calf Thymus," *JEM* 99 (1954): 65–78; James G. Hirsch, "Phagocytin: A Bactericidal Substance from Polymorphonuclear Leucocytes," and "The Studies of the Bactericidal Action of Phagocytin," *JEM* 103 (1956): 589–611, and 613–621.

38. Paul B. Beeson, "Infectious Diseases (Microbiology)," in *Advances in American Medicine: Essays at the Bicentennial*, edited by John Z. Bowers and Elizabeth F. Purcell (New York: Josiah Macy, Jr. Foundation, 1976), pp. 100–156, quotation on p. 128.

39. Gardner Middlebrook and Diran Yegian, "Certain Effects of Streptomycin on Mycobacteria *in vitro*," *ART* 54 (1946): 553–558; Margaret A. Jennings, Howard W. Florey, and Norman G. Heatley, "Acquired Resistance of Micro-organisms to Streptomycin and the Action of Streptomycin," in Howard W. Florey, Ernst Chain, Norman G. Heatley, Margaret A. Jennings, A. Gordon Sanders, Edward P. Abraham, and M. Ethel Florey, *Antibiotics: A Survey of Penicillin, Streptomycin, and Other Antimicrobial Substances from Fungi, Actinomycetes, Bacteria, and Plants* (London: Oxford University Press, 1949), pp. 1365–1400.

40. James G. Hirsch, Russell W. Schaedler, Cynthia H. Pierce, and I. Maclean Smith, "A Study Comparing the Effects of Bed Rest and Physical Activity on Recovery from Pulmonary Tuberculosis," *ART* 75 (1957): 359–367.

41. Cynthia H. Pierce, James G. Hirsch, and Russell W. Schaedler, "Rapid Detection of Drug-Resistant Tubercle Bacilli in Sputum by Slide Cultures," *ART* 75 (1957): 331–337.

42. RD, "Anti-Infectious Agents of Natural Origin," in *Medicine in the Postwar World* (New York: Columbia University Press, 1948), pp. 92–106.

43. Rollin D. Hotchkiss, "From Microbes to Medicine: Gramicidin, René Dubos, and the Rockefeller," in Moberg and Cohn, *Launching the Antibiotic Era*, p. 16.

44. Davis, "Two Perspectives," p. 72.

45. Arthur Conan Doyle, "Dr. Koch and his Cure," *Review of Reviews* 1 (1890): 552.

46. Diran Yegian and Keith R. Porter, "Some Artifacts Encountered in Stained Preparations of Tubercle Bacilli. I. Non-Acid-Fast Forms Arising from Mechanical Treatment," *Journal of Bacteriology* 48 (1944): 83–91; and idem, "Some Artifacts Encountered in Stained Preparations of Tubercle Bacilli. II. Much Granules and Beads," *Journal of Bacteriology* 50 (1945): 563–575.

47. OH, Part II, p. 14 (14, 15).

48. Rollin D. Hotchkiss, "The Nature of the Bactericidal Action of Surface Active Agents," *Annals of the New York Academy of Sciences* 46 (1946): 479–492.

49. RD and Bernard D. Davis, "Factors Affecting the Growth of Tubercle Bacilli in Liquid Media," *JEM* 83 (1946): 409–423; idem, "The Binding of Fatty Acids by Serum Albumin, a Protective Growth Factor in Bacteriological Media," *JEM* 86 (1947): 215–228: RD and Gardner Middlebrook, "Media for Tubercle Bacilli," *ART* 56 (1947): 334–345.

50. RD, Bernard D. Davis, Gardner Middlebrook, and Cynthia Pierce, "The Effect of Water Soluble Lipids on the Growth and Biological Properties of Tubercle Bacilli," *ART* 54 (1946): 204–212; and RD and Gardner Middlebrook, "The Effect of Wetting Agents on the Growth of Tubercle Bacilli," *JEM* 88 (1948): 81–88.

51. RD, "The Effect of Lipids and Serum Albumin on Bacterial Growth," *JEM* 85 (1947): 9–22; and idem, "The Effect of Sphingomyelin on the Growth of Tubercle Bacilli," *JEM* 88 (1948): 73–79.

52. Pierce, Hirsch, and Schaedler, "Rapid Detection."

53. RD, Frank Fenner, and Cynthia H. Pierce, "Properties of a Culture of BCG Grown in Liquid Media Containing Tween 80 and the Filtrate of Heated Serum," *ART* 61 (1950): 66–76; and RD and Frank Fenner, "Production of BCG Vaccine in a Liquid Medium Containing Tween 80 and a Soluble Fraction of Heated Human Serum. I. Production and Viability of the Culture," and "II. Antigenicity of the Culture after Various Periods of Storage," *JEM* 91 (1950): 261–284.

54. Gardner Middlebrook, RD, and Cynthia Pierce, "Virulence and Morphological Characteristics of Mammalian Tubercle Bacilli," *JEM* 86 (1947): 175–184.

55. Cynthia Pierce, RD, and Gardner Middlebrook, "Infection of Mice with Mammalian Tubercle Bacilli Grown in Tween-Albumin Liquid Medium," *JEM* 86 (1947): 159–174.

56. RD, "The Experimental Analysis of Tuberculous Infections," *Experientia* (Basel) 3 (1947): 45–52.

57. Herbert S. Gasser, "Report of the Director of the Institute to the Corporation," *BSD* 33 (October 1945): 8.

58. "Progress Report," *Time*, 24 June 1946, p. 61; and RD, "Progress Exaggerated" [Letter to the Editor], *Time*, 27 July 1946, p. 10. As a result of RD's agitation over this report, he met Marian MacPhail, director of research at Time, Inc., who introduced him to her husband, Walsh McDermott, a tuberculosis specialist at Cornell University Medical College. The happy conclusion was that the two men became very close personal friends.

59. RD, *The Bacterial Cell*, p. 193.

60. RD, "Properties and Structures of Tubercle Bacilli Concerned in Their Pathogenicity," in *Mechanisms of Microbial Pathogenicity*, edited by J.W. Howie and A.J. O'Hea (Cambridge: University Press, 1955), pp. 103–125.

61. William R. Jacobs, Jr. and Barry R. Bloom, "Molecular Genetic Strategies for Identifying Virulence Determinants of *Mycobacterium tuberculosis*," in *Tuberculosis: Pathogenesis, Protection, and Control*, edited by Barry R. Bloom (Washington, D.C.: ASM Press, 1994), pp. 253–268.

62. M.J. Mahan, D.M. Heithoff, R.L. Sinsheimer, and D.A. Low, "Assessment of Bacterial Pathogenesis by Analysis of Gene Expression in the Host," *Annual Review of Genetics* 34 (2000): 139–164.

63. Joshua Lederberg, "Crowded at the Summit: The Future of Infectious Disease," unpublished Rockefeller University Centennial Lecture on Science and Society, New York, 26 January 2001.

64. Dubos and Dubos, *White Plague*, p. 122.

65. Frank Fenner, Samuel P. Martin, and Cynthia H. Pierce, "The Enumeration of Viable Tubercle Bacilli in Cultures and in Infected Tissues," *Annals of the New York Academy of Sciences* 52 (1949): 751–764; Cynthia H. Pierce, RD, and Gardner Middlebrook, "Infection of Mice with Mammalian Tubercle Bacilli"; Samuel P. Martin, Cynthia H. Pierce, Gardner Middlebrook, and RD, "The Effect of Tubercle Bacilli on the Polymorphonuclear Leucocytes of Normal Animals," *JEM* 91 (1950): 381–392; Cynthia H. Pierce, RD, and Werner B. Schaefer, "Multiplication and Survival of Tubercle Bacilli in the Organs of Mice," *JEM* 97 (1953): 189–206; and idem, "Differential Characteristics in Vitro and in Vivo of Several Substrains of BCG. I. Multiplication and Survival in Vitro," "II. Morphologic Characteristics in Vitro and in Vivo," "III. Multiplication and Survival in Vivo," and "IV. Immunizing Effectiveness," *ART* 74 (1956): 655–666, 667–682, 683–698, and 699–717.

66. Middlebrook, RD, and Pierce, "Virulence and Morphological Characteristics."

67. Hubert Bloch, "Studies on the Virulence of Tubercle Bacilli. Isolation and Biological Properties of a Constituent of Virulent Organisms," *JEM* 91 (1950): 197–218; H. Noll, H. Bloch, J. Asselineau, and E. Lederer, "The Chemical Structure of

the Cord Factor of *Mycobacterium tuberculosis*," *Biochimica et Biophysica Acta* 20 (1956): 299–309; and G.S. Besra and D. Chatterjee, "Lipids and Carbohydrates of *Mycobacterium tuberculosis*," in *Tuberculosis*, ed. Bloom, pp. 285–306.

68. RD and Gardner Middlebrook, "Cytochemical Reaction of Virulent Tubercle Bacilli," *ART* 58 (1948): 698–699; and RD and Emanuel Suter, "The Effect of Ammonium Ions and Aliphatic Amines on the Ability of Virulent Mycobacteria to Bind Neutral Red," *ART* 60 (1949): 384.

69. RD, "Cellular Structures and Functions Concerned in Parasitism," *Bacteriological Reviews* 12 (1948): 173–194; and RD, "Tuberculosis," *Scientific American* 181 (1949): 30–41.

70. Gardner Middlebrook and RD, "Specific Serum Agglutination of Erythrocytes Sensitized with Extracts of Tubercle Bacilli," *JEM* 88 (1948): 521–528; and Gardner Middlebrook, "The Hemagglutination Test in Tuberculosis," *ART* 62 (1950): 223–226.

71. Arne Lind and Malin Ridell, "Immunologically Based Diagnostic Tests: Humoral Antibody Methods," in *The Mycobacteria: A Sourcebook*, edited by George P. Kubica and Lawrence G. Wayne (New York: Marcel Dekker, Inc., 1984), pp. 221–248.

72. RD, *Biochemical Determinants.*

73. Albert Calmette, Camille Guérin, Auguste Boquet, and Léopold Nègre, *La vaccination préventive contre la tuberculose par le 'BCG'* (Paris: Masson et Cie., 1927); and Camille Guérin, "The History of BCG," in *BCG Vaccination against Tuberculosis*, edited by S.R. Rosenthal (Boston: Little, Brown, 1957), pp. 38–43.

74. RD, "Immunological Aspects of BCG Vaccination," *ART* 60 (1949): 670–674.

75. RD, Fenner, and Pierce, "Properties of a Culture of BCG"; and RD and Fenner, "Production of BCG Vaccine."

76. RD, *The Bacterial Cell*, pp. 232–233; and Louis Pasteur, "De l'attenuation des virus et de leur retour à la virulence," *Comptes Rendus des Séances de l'Académie des Sciences* 92 (1881): 429–435.

77. W. Emanuel Suter and RD, "Variability of BCG Strains (Bacillus Calmette–Guérin)," *JEM* 93 (1951): 559–572; and RD, Cynthia H. Pierce, and Werner B. Schaefer, "Antituberculous Immunity Induced in Mice by Vaccination with Living Cultures of Attenuated Tubercle Bacilli," *JEM* 97 (1953): 207–220.

78. RD and Robert Debré, Introduction to "Centre International de l'Enfance sur les critères d'activité du BCG," *Revue d'Immunologie* (Paris) 19 (1955): 117–120; G. Conge, E. Collin, F.M. Lévy, and RD, "Microscopic Enumeration of Mycobacteria in Pulmonary Lesions of Tuberculous Mice. Effect of Prior BCG Vaccination,"

ART 79 (1959): 484–491; *Methods for the Study of BCG Vaccine*, a special issue of *Bulletin of the International Union Against Tuberculosis*, 1960; Maurice Panisset, editor, *Méthodes expérimentales d'étude du vaccin BCG* (Montreal: Thérien Frères, 1963); and F.M. Lévy, *Recherches sur le BCG. Techniques d'étude et problèmes de standardisation* (Paris: Flammarion, 1966).

79. RD, Herman Hilleboe, Horace L. Hodes, Walsh McDermott, Gardner Middlebrook, Rufus Payne, James E. Perkins, Leon H. Schmidt, Jacob Yerushalmy, and Esmond R. Long, "Report of Ad Hoc Advisory Committee on BCG to the Surgeon General of the United States Public Health Service," *ART* 76 (1957): 726–731.

80. Léopold Nègre and Auguste Boquet, *Antigénothérapie de la tuberculose par les extraits methyliques de bacilles de Koch* (Paris: Masson, 1927); RD, Werner B. Schaefer, and Cynthia H. Pierce, "Antituberculous Immunity in Mice Vaccinated with Killed Tubercle Bacilli," *JEM* 97 (1953): 221–233; David W. Weiss and RD, "Antituberculous Immunity Induced in Mice by Vaccination with Killed Tubercle Bacilli or with a Soluble Bacillary Extract," *JEM* 101 (1955): 313–330; and idem, "Antituberculous Immunity Induced by Methanol Extracts of Tubercle Bacilli—Its Enhancement by Adjuvants," *JEM* 103 (1956): 73–85.

81. RD to John Adair, 28 March 1980, RD Papers.

82. RD, "Man Meets his Environment," in *Health and Nutrition. Science, Technology and Development*, volume six, U.S. Papers prepared by the International Development Agency, U.S. State Department, for the United Nations Conference "Application of Science and Technology for the Benefit of the Less Developed Areas" (Washington, D.C.: United States Government Printing Office Document 5096, 1963), pp. 1–11; and David S. Jones, *Rationalizing Epidemics: Meanings and Uses of American Indian Mortality since 1600* (Cambridge: Harvard University Press, 2004).

83. Walsh McDermott to Paul Sears, quoted in John Adair and Kurt W. Deuschle, *The People's Health: Medicine and Anthropology in a Navajo Community* (New York: Appleton Century Crofts, 1970), p. 31. This monograph is the final report of the Navajo Field Project.

84. "Immunizing Antigen of Tubercle Bacilli and Cornell University Medical School (Walsh McDermott)," RD Papers, Box 4, Folder 5; Walsh McDermott, M.D., Papers, 1922–1982, New York Weill Cornell Medical Center Archives, New York, New York; and Jean Dubos, interview with the author.

85. RD, "The Problem of Tuberculosis," in *Report of the Secretary General on Establishing United Nations Research Laboratories* (United Nations Economic and Social Council Document E/620, 23 January 1948), pp. 270–287; and RD, "Acquired Immunity to Tuberculosis," *ART* 90 (1964): 505–515.

86. Barry Bloom and Paul E.M. Fine, "The BCG Experience: Implications for Future Vaccines against Tuberculosis," in *Tuberculosis*, ed. Bloom, pp. 537, 539.

87. M.A. Horwitz, B.-W. E. Lee, B.J. Dillon, and G. Harth, "Protective Immunity against Tuberculosis Induced by Vaccination with Major Extracellular Proteins of *Mycobacterium tuberculosis*," *Proceedings of the National Academy of Sciences U.S.A.* 92 (1995): 1530–1534.

88. World Health Organization Tuberculosis Programme, *TB: A Global Emergency* (Geneva: World Health Organization, 1994); Thomas A. Shinnick, "Tuberculosis—Present and Future," in *Microbe Hunters—Then and Now*, edited by Hilary Koprowski and Michael B.A. Oldstone (Bloomington, Illinois: Medi-Ed Press, 1996), pp. 319–328; John D. McKinney, William R., Jacobs, Jr., and Barry R. Bloom, "Persisting Problems in Tuberculosis," in *Emerging Infections*, edited by Richard Krause (San Diego: Academic Press, 1998), pp. 51–146; and Laurie Garrett, *Betrayal of Trust: The Collapse of Global Public Health* (New York: Hyperion, 2000).

89. RD, "The Problem of Tuberculosis."

90. RD to Jacques Monod, 19 September 1951, Monod Papers, Archives of the Pasteur Institute, Paris. At this time, Dubos also turned over the writing of the chapter on morphology and physiology of bacteria for the second edition of *Bacterial and Mycotic Infections of Man* to Monod and A.M. Pappenheimer, who introduced a section on bacterial genetics, a topic remote from Dubos' interests.

Chapter 4 Mirage of Health: Infection versus Disease

1. RD to Adéline Dubos, 1944, RD Family Papers.

2. RD to Adéline Dubos, 1946, RD Family Papers.

3. RD to Adéline Dubos, n.d. [1953], RD Family Papers.

4. Jean Dubos to Henriette Noufflard Guy-Loë, 31 October 1951, author's papers.

5. RD to Madeleine Vorwald, 6 January 1977, RD Papers.

6. RD, "How to Use More Personal Energy to Live Longer, Feel Better, and Enjoy Life More," interview with Beverly Russell, *House & Garden*, January 1974, pp. 16, 18.

7. George W. Corner, *A History of The Rockefeller Institute, 1901–1953: Origins and Growth* (New York: The Rockefeller Institute Press, 1964), p. 328; Herbert S. Gasser, "Medical Research: A Look Ahead at The Rockefeller Institute for Medical Research," *BSD* 39 (26 October 1951): 1–94; Merrill W. Chase and Carlton C. Hunt, "Herbert Spencer Gasser, 1888–1963," *Biographical Memoirs* 67 (1995): 146–177.

8. RD, "Centenary Tribute to Peyton Rous," *JEM* 150 (1979): 737–739; and Peyton Rous in *Memorial Meeting for Simon Flexner* (Boston: The Merrymount Press, 12 June 1946), p. 17.

9. Around 1960, RD decided *The Journal of Experimental Medicine* was neglecting the field of pathogenesis, which his laboratory was studying. According to Rous,

RD became "irked by the need, forced on us, to decline desirable papers" on this topic and started discussions with prominent physician-scientists to create a journal devoted to pathogenesis. Rous greeted the plan "warmly as relieving a bad situation." While the plan was taking shape, RD moved into another new field he called environmental biomedicine and abandoned all efforts to found a journal devoted to his previous interests. Peyton Rous Papers, American Philosophical Society, Philadelphia, Pennsylvania.

10. RD, "Cellular Structures and Functions Concerned in Parasitism," *Bacteriological Reviews* 12 (1948): 173–194, quotations are on pp. 176, 189.

11. OH, Part II, p. 137 (678).

Bernard Davis, a member of Dubos' laboratory, contributed two chapters—"Principles of Sterilization" and "Principles of Chemotherapy"—to *Bacterial and Mycotic Infections of Man*, pp. 637–682. Davis asserted that further promising antibiotics would continue to appear, especially with the biochemical knowledge emerging from the new science of bacterial genetics. At the time he believed his lab chief had become too "bearish on antibiotics" but years later granted that Dubos' view "might have helped to dispel the growing, widespread misconception that antibiotics would solve all problems of infectious disease." Bernard D. Davis, "Two Perspectives: On René Dubos, and on Antibiotic Actions," in *Launching the Antibiotic Era*, edited by Carol L. Moberg and Zanvil A. Cohn (New York: Rockefeller University Press, 1990), p. 72.

12. RD, "The Living Thought of Louis Pasteur," *MD* 7 (1952): 381–384, 402.

13. RD, *Louis Pasteur*, p. 83.

14. RD to Adéline Dubos, 1948, RD Family Papers.

15. OH, Part II, p. 2 (2).

16. RD, handwritten notes on reading this debate, 1944, RD Papers, Box 78, Folder 7.

17. RD, *Louis Pasteur*, p. 387.

18. RD, "Background and Prejudices," *Dreams of Reason*, pp. 1–11.

19. RD Papers, Box 78, Folder 8.

20. RD, *Louis Pasteur*, pp. 23, 88, 89.

21. Ibid., p. 266.

22. Ibid., pp. 231, 287.

23. Ibid., p. 273.

24. Ibid., p. 279.

25. Ibid., pp. 280–282.

26. Ibid., p. 290.

27. Ibid., pp. 379–381.

28. The biography's publication history set a general precedent for RD's subsequent books; they continued to sell year after year, were translated into several languages, reprinted many times, and remained in print for decades. Of the more than one hundred biographies of Pasteur, the Dubos work was singled out in 1993 by Annick Perrot, *conservateur* of the Pasteur Institute Archives, as the only one to understand and evaluate the entire body of Pasteur's science, interview with the author, Paris, June 1993.

29. RD, *Louis Pasteur*, p. 187.

30. RD, "The Complexities of Genius," a new introduction to *Louis Pasteur: Free Lance of Science* (New York: Charles Scribner's Sons, 1976), pp. xxvii–xxxix.

31. Arnold R. Rich, *The Pathogenesis of Tuberculosis* (Springfield, Illinois: Charles C. Thomas, 1944), pp. 325–326.

32. RD, "Tuberculosis," *Scientific American* 181 (1949): 30–41.

33. RD, "The Experimental Analysis of Tuberculous Infections," *Experientia* (Basel) 3 (1947): 45–52.

34. OH, Part II, pp. 34–36 (128–130).

35. Cynthia Pierce, RD, and Gardner Middlebrook, "Infection of Mice with Mammalian Tubercle Bacilli Grown in Tween-Albumin Liquid Medium," *JEM* 86 (1947): 159–174; Ian M. Orme and Frank M. Collins, "Mouse Model of Tuberculosis," in *Tuberculosis: Pathogenesis, Protection, and Control*, edited by Barry R. Bloom (Washington, D.C.: ASM Press, 1994), pp. 113–134.

The versatile C57 Black mouse, one of science's most popular and best documented strains, served as the reference strain for the sequencing of the mouse genome.

36. RD, *Biochemical Determinants*; RD, "Biochemical Determinants of Infection," *Bulletin of the New York Academy of Medicine* 31 (1955): 5–19; RD, "The Microenvironment of Inflammation or Metchnikoff Revisited," *The Lancet* ii (1955): 1–5; and RD, "Metabolic Interrelationships between Host and Parasite," in *Host-Parasite Relationships in Living Cells*, edited by Harriet M. Felton (Springfield, Illinois: Charles C. Thomas, 1957), pp. 172–190.

37. RD, "The Effect of Organic Acids on Mammalian Tubercle Bacilli," *JEM* 92 (1950): 319–332.

38. RD, "Viability of Tubercle Bacilli in Vivo with and without Chemotherapy," *ART* 67 (1953): 874–877; RD, "Effect of Ketone Bodies and Other Metabolites on the Survival and Multiplication of Staphylococci and Tubercle Bacilli," *JEM* 98 (1953): 145–155.

39. RD, *Biochemical Determinants.*

40. RD, "The Unknowns of Staphylococcal Infection," *Annals of the New York Academy of Sciences* 65 (1956): 243–246.

41. RD, "Retrospectives and Prospectives," in *Proceedings of the First International Conference on the Use of Antibiotics in Agriculture* (Washington, D.C.: National Academy of Sciences-National Research Council, Publication 397, 1956), pp. xv–xx.

42. Henry Welch, "Antibiotics 1943–1955: Their Development and Role in Present-Day Society," and Herbert G. Luther, "The Application of Antibiotics to the Livestock Industry," in *The Impact of the Antibiotics on Medicine and Society,* edited by Iago Galdston (New York: International Universities Press, 1958), pp. 70–87 and 121–163.

43. Quoted in Lennard Bickel, *Rise Up to Life: A Life of Howard Walter Florey Who Gave Penicillin to the World* (London: Angus and Robertson, 1972), p. 253.

44. RD, "Retrospectives."

45. Marc Lappé, *Germs That Won't Die: Medical Consequences of the Misuse of Antibiotics* (Garden City, New York: Anchor Press/Doubleday, 1982); Stuart B. Levy, *The Antibiotic Paradox: How Miracle Drugs are Destroying the Miracle* (New York: Plenum Press, 1992); and Laurie Garrett, "The Revenge of the Germs, or Just Keep Inventing New Drugs," in *The Coming Plague: Newly Emerging Diseases in a World Out of Balance* (New York: Farrar, Straus and Giroux, 1994), pp. 441–456.

46. Harold J. Simon, *Attenuated Infection: The Germ Theory in Contemporary Perspective* (Philadelphia: Lippincott, 1960); and Theodor Rosebury, *Microorganisms Indigenous to Man* (New York: McGraw-Hill, 1962).

47. RD, *Biochemical Determinants,* p. 2.

48. Joseph C. Aub, Austin M. Brues, RD, Seymour S. Kety, Ira T. Nathanson, Alfred Pope, and Paul C. Zamecnik, "Bacteria and the 'Toxic Factor' in Shock," *War Medicine* 5 (1944): 71–73; and Alfred Pope, Paul C. Zamecnik, Joseph C. Aub, Austin M. Brues, RD, Ira T. Nathanson, and A.L. Nutt, "The Toxic Factors in Experimental Traumatic Shock. VI. The Toxic Influence of the Bacterial Flora, particularly *Clostridium welchii,* in Exudates of Ischemic Muscle," *Journal of Clinical Investigation* 24 (1945): 856–863.

49. RD, "The Host in Tuberculosis," *Acta Tuberculosea Scandinavica* 37 (1959): 42–60.

50. Mogens Volkert, Cynthia Pierce, Frank L. Horsfall, Jr., and RD, "The Enhancing Effect of Concurrent Infection with Pneumotropic Viruses on Pulmonary Tuberculosis in Mice," *JEM* 86 (1947): 203–214.

51. RD, "Effect of Ketone Bodies"; and idem, "Effect of Metabolic Factors on the Susceptibility of Albino Mice to Experimental Tuberculosis," *JEM* 101 (1955): 59–84.

52. Dubos and Dubos, *The White Plague.*

53. Peyton Rous to Herbert S. Gasser, 20 June 1946, Peyton Rous Papers, American Philosophical Society, Philadelphia, Pennsylvania.

54. RD and Cynthia Pierce, "The Effect of Diet on Experimental Tuberculosis of Mice," *ART* 57 (1948): 287–293.

55. J. Maclean Smith and RD, "The Effect of Nutritional Disturbances on the Susceptibility of Mice to Staphylococcal Infections," *JEM* 103 (1956): 109–118.

56. Russell W. Schaedler and RD, "Reversible Changes in the Susceptibility of Mice to Bacterial Infections. II. Changes Brought about by Nutritional Disturbances," *JEM* 104 (1956): 67–84.

57. J. Maclean Smith and RD, "The Effect of Dinitrophenol and Thyroxin on the Susceptibility of Mice to Staphylococcal Infections," *JEM* 103 (1956): 119–126.

58. RD and Russell W. Schaedler, "Effect of Dietary Proteins and Amino Acids on the Susceptibility of Mice to Bacterial Infections," *JEM* 108 (1958): 69–81 and *JEM* 110 (1959): 921–934.

59. RD and Russell W. Schaedler, "Reversible Changes in the Susceptibility of Mice to Bacterial Infections. I. Changes Brought about by Injection of Pertussis Vaccine or of Bacterial Endotoxins," *JEM* 104 (1956): 53–65; Samuel J. Prigal and RD, "Effect of Allergic Shock on Fate of Staphylococci in the Organs of Mice," *Proceedings of the Society for Experimental Biology and Medicine* 93 (1956): 340–343; and RD and Russell W. Schaedler, "Effects of Cellular Constituents of Mycobacteria on the Resistance of Mice to Heterologous Infections. I. Protective Effects" and " II. Enhancement of Infection," *JEM* 106 (1957): 703–726.

60. Charles LeMaistre and Ralph Tompsett, "The Emergence of Pseudotuberculosis in Rats Given Cortisone," *JEM* 95 (1952): 393–408; J. Seronde, Jr., "The Resistance of Normal Rats to Inoculation with a *Corynebacterium* Pathogenic in Pantothenate Deficiency," *Proceedings of the Society for Experimental Biology and Medicine* 85 (1954): 521–524; and I.L. Schechmeister and F.L. Adler, "Activation of Pseudotuberculosis in Mice Exposed to Sublethal Total Body Radiation," *Journal of Infectious Diseases* 92 (1953): 228–239.

61. Cynthia H. Pierce-Chase, Robert M. Fauve, and RD, "Corynebacterial Pseudotuberculosis in Mice. I. Comparative Susceptibility of Mouse Strains to Experimental Infection with *Corynebacterium kutscheri*" and "II. Activation of Natural and Experimental Latent Infections," *JEM* 120 (1964): 267–304.

62. RD, *Man Adapting*, p. 415.

63. RD, "Utilization of Selective Microbial Agents in the Study of Biological Problems," *The Harvey Lectures* 35 (1939–1940): 223–242.

64. RD, "Second Thoughts on the Germ Theory," *Scientific American* 192 (May 1955): 31–35.

65. RD, "Les variations de la susceptibilité à l'infection," in *Aux frontières de la microbiologie médicale*, edited by Robert Fasquelle (Paris: Éditions Médicales Flammarion, 1961), pp. 49–63.

66. RD, *Louis Pasteur*, p. 227.

67. RD to Adéline Dubos, 1949, RD Family Papers.

68. "Flair Personified: Dr. René Dubos," *Flair*, July 1950, pp. 20–23.

69. RD, interview with Lewis Nichols, *The New York Times Book Review*, 21 December 1952.

70. Dubos and Dubos, *The White Plague*, p. 156.

71. RD, "The Philosopher's Search for Health," *Transactions of the Association of American Physicians* 66 (1953): 31–41.

72. RD to David Rockefeller, 21 April 1952, RD Papers.

73. OH, Part II, pp. 145–156 (679–689).

74. OH, Part II, p. 120 (574).

75. Selman A. Waksman, "TB, There is the Enemy," *The New York Times Book Review*, 16 November 1952.

76. From all accounts, the relationship between RD and Waksman was always respectful. In the 1955 Oral History, RD said they were "good friends." Clearly, the two led different scientific lives and had different temperaments. As compared by colleagues Rollin Hotchkiss and Bernard Davis in *Launching the Antibiotic Era*, Waksman was a pragmatic, industrious microbiologist, whereas RD was a restless scientist creating new approaches to problems. In 1951, Waksman asked RD and Harry Eagle to be cofounders with him of the Foundation for Microbiology that was created with royalties from streptomycin. RD remained a trustee until 1959. In 1954, he also gave the keynote speech at the dedication of Waksman's Institute of Microbiology (now The Waksman Institute). Waksman served as its president for four years and in 1958 asked RD to be his successor, an invitation that was declined. Selman Waksman Papers, Special Collections and Archives, Rutgers University Archives, New Brunswick, New Jersey.

77. RD to Adéline Dubos, 1 January 1950, RD Family Papers.

78. RD, "Philosopher's Search for Health," p. 38.

79. RD, "Microbiology in Fable and Art," *Bacteriological Reviews* 16 (1952): 145–151; RD, "The Gold-Headed Cane in the Laboratory," in *National Institutes of*

Health Annual Lectures (Washington, D.C.: United States Government Printing Office, Public Health Service Publication Number 388, 1953), pp. 89–102; RD, "The Micro-environment of Inflammation"; RD, "Second Thoughts on the Germ Theory"; and RD, "The Evolution and Ecology of Microbial Diseases," in *Bacterial and Mycotic Infections of Man*, third edition, 1958, pp. 14–27.

80. RD, "Pasteur in His Time," in *The Pasteur Fermentation Centennial, 1857–1957*, edited by John E. McKeen (New York: Charles Pfizer & Company, 1958), pp. 188–193; quotation on p. 191.

81. RD, "The Evolution of Infectious Diseases in the Course of History," *The Canadian Medical Association Journal* 79 (1958): 445–451; and RD, *Man Adapting*, pp. 426 and 431.

82. RD, *Biochemical Determinants*.

83. Julian Huxley to Cass Canfield, 7 January 1959, and RD to John Appleton, 13 January 1959, RD Papers.

84. RD, "Medical Utopias," *Daedalus* 88 (1959): 410–424.

85. RD, "Evolution of Infectious Diseases."

86. RD, *Mirage of Health*, p. 233.

87. Ibid., p. 235.

88. Ibid., pp. 136, 132.

89. Walsh McDermott to Rhoda Schlamm, 22 February 1980, Walsh McDermott, M.D., Papers, 1922–1982, New York Weill Cornell Medical Center Archives, New York, New York, Box 27, Folder 11.

90. Daniel Callahan, *False Hopes: Overcoming the Obstacles to a Sustainable, Affordable Medicine* (New Brunswick: Rutgers University Press, 1999), pp. 65, 246; Marc Lappé, *Germs That Won't Die*; Marc Lappé, *Evolutionary Medicine: Rethinking the Origins of Disease* (San Francisco: Sierra Club Books, 1994); Randolph M. Nesse and George C. Williams, *Why We Get Sick: The New Science of Darwinian Medicine* (New York: Times Books, 1994); Paul W. Ewald, *Evolution of Infectious Disease* (New York: Oxford University Press, 1994); and Anthony J. McMichael, *Human Frontiers, Environments, and Disease: Past Patterns, Uncertain Futures* (Cambridge: University Press, 2001).

91. RD, "Science and Conscience in Modern Medicine," *Journal of the American Osteopathic Association* 60 (1960): A187–195.

92. RD, "The Future of Infectious Diseases," *Comprehensive Therapy* 1 (1975): 15–18.

Chapter 5 Toward a Science of Human Nature

1. The publication of *The Torch of Life* coincided with the longest newspaper strike in New York City.

2. Ilya Prigogine, "L'ordre à partir du chaos?" *Prospective et Santé* 13 (1980): 29–58. Translated from the original French.

3. RD to James Parks Morton, Dean of the Cathedral of St. John the Divine, 16 May 1975, RD Papers. Although not a member of this Episcopalian church, RD gave at least five sermon-dialogues in the late 1970s at the request of Dean Morton. Also, along with several prominent ecologists, RD was made a Cathedral Colleague in 1975. Following Dubos' death, Dean Morton engaged Spanish architect Santiago Calatrava to design the René Dubos Bioshelter that would become the south transept of the Cathedral; this transept remains an architectural model, once exhibited at the Museum of Modern Art in New York, that has yet to be built. John Brodie, "A Garden of Eden for St. John's," *The New York Times*, 26 January 1992.

4. RD to Wilfred L. Morin, 8 April 1980, and RD to Pat Reif, 8 July 1970, RD Papers; and RD to Henriette Noufflard, 30 Nov 1955, author's papers.

5. RD to Walsh McDermott, 10 May 1977, RD Papers. RD was married in New York City churches selected by his wives, Marie Louise Bonnet in a Roman Catholic church and Jean Porter in an Episcopalian church.

6. RD, "Homo Sapiens: Genus or Sentient,"*New York State Journal of Medicine* 81 (1981): 1872–1875; and Robert Serrou, "Un grand savant français parle/René Dubos," *Paris Match* (4 April 1980): 3–10. Translated from the original French.

7. RD, *Louis Pasteur*, p. 185.

8. André Lwoff to RD, 23 November 1965, RD Papers. Translated from the original French.

9. RD, *Dreams of Reason*, p. 125.

10. The ideas of RD presented here are far removed from a very recent proposal for "sciences of human nature" that are based on efforts to explain mind and behavior as products of natural selection and genetic endowment, for example, in Steven Pinker, *The Blank Slate: The Modern Denial of Human Nature* (New York: Viking, 2002).

11. RD, *The Torch of Life*, p. 106.

12. RD, *Man Adapting*, p. 332.

13. Howard A. Schneider, "Experimental Medical Ecology," *Journal of the American Medical Association* 194 (1965): 157–159.

14. RD, "Hippocrates in Modern Dress," *Perspectives in Biology and Medicine* 9 (1966): 275–288.

15. Surprisingly, there are few published criticisms, reports of conflict, or instances of personal antagonism toward Dubos and his work. Rollin Hotchkiss, Dubos' colleague from 1935 until he retired, describes his whole career as free of any serious enemies or conflict. He attributes this to Dubos' ability to anticipate "questioning inquiries or doubts" of others and to practice "well the ancient dictum of 'being your own severest critic'. . . . He escaped much criticism by doing it himself."

Even with almost twenty years of seeking criticism and conflict in Dubos' life, I have found nothing to contradict what Hotchkiss observed. Almost without exception, the several hundred book reviews, scientific journal reports, newspaper articles, print and film interviews found while researching his bibliography and this biography were neutral or favorable.

A similar observation was reported in a dissertation on Dubos by Mark F. Dreessen who offered the caveat that the critical reviewers "either reported his ideas without comment or supported them. This has presented a difficult task for one trying to present a critical assessment since the impression might easily be conveyed that critical reviews have not been considered or have been omitted." Mark F. Dreessen, "René Dubos and the Public Understanding of Science" (Ph.D. dissertation, New York University, 1985).

A few speculations on why this is so are worthwhile. Nothing has been found to show that Dubos engaged in personal attacks on individuals or organizations. Nor did Hotchkiss "observe any attempt [by Dubos] to initiate trouble for others," adding that "the closest to an 'enemy' I knew of was his own co-worker, Bernie Davis—by then at Harvard—who had a distaste for the style of Dubos' public presentations. . . . But it was surely the love-hate of a rebellious growing youngster."

Apparently, there was no reason for anyone to attack his science until Dubos left the tuberculosis work. After that, his scientific explorations were clearly explained as speculative and personal with aspects of common sense appeal, even if this approach was unconventional for a scientist. His arguments about health and environment were based on enlightened self-interest, which reverberated in his audiences about personal awareness, responsibility, and stewardship. Who would argue with that in public? Finally, the few criticisms from writers in the environmental/conservation/preservation bloc tend to come from the authors' specialized viewpoints, and they are phrased more as points in a debate than as bruising or hostile attacks against Dubos himself; their observations are discussed in chapter 6.

The comments of Hirsch, Cohn, and Hotchkiss are personal communications with the author.

16. Herbert S. Gasser, "Medical Research: A Look Ahead," *BSD* 39 (26 October 1951): 1–94.

17. RD and Russell W. Schaedler, "The Digestive Tract as an Ecosystem," *American Journal of the Medical Sciences* 248 (1964): 267–271.

18. Dwayne C. Savage and RD, "Localization of Indigenous Yeast in the Murine Stomach," *Journal of Bacteriology* 94 (1967): 1811–1816; and Dwayne C. Savage, RD, and Russell W. Schaedler, "The Gastrointestinal Epithelium and Its Autochthonous Bacterial Flora," *JEM* 127 (1968): 67–76.

19. Russell W. Schaedler, RD, and Richard Costello, "The Development of the Bacterial Flora in the Gastrointestinal Tract of Mice," *JEM* 122 (1965): 59–66; RD, Russell W. Schaedler, Richard Costello, and Philippe Hoet, "Indigenous, Normal, and Autochthonous Flora of the Gastrointestinal Tract," *JEM* 122 (1965): 67–76; James H. Gordon and RD, "The Anaerobic Bacterial Flora of the Mouse Cecum," *JEM* 132 (1970): 251–260; Adrian Lee, James Gordon, Chi-Jen Lee, and RD, "The Mouse Intestinal Microflora with Emphasis on the Strict Anaerobes," *JEM* 133 (1971): 339–352; and Dwayne C. Savage, "Microbial Biota of the Human Intestine: A Tribute to Some Pioneering Scientists," *Current Issues in Intestinal Microbiology* 2 (2001): 1–15.

20. RD, "The Gastrointestinal Microbiota of the So-Called Normal Mouse," *Carworth Europe Collected Papers* 2 (1968): 11–18.

21. After leaving the laboratory, Dwayne Savage and Adrian Lee continued studies to identify other gastrointestinal microbes. Dwayne C. Savage, "Microbial Ecology of the Gastrointestinal Tract," *Annual Review of Microbiology* 31 (1977): 107–133; and Adrian Lee and J. O'Rourke, "Gastric Bacteria Other than *Helicobacter pylori*," *Gastroenterology Clinics of North America* 22 (1993): 21–42.

Today, there is renewed interest in beneficial microbial associations in the digestive tract. A discipline devoted to *probiotics* has its roots in treatments at the Pasteur Institute in the early 1900s when Nobel laureate Élie Metchnikoff administered therapeutic microbes to patients with intestinal diseases. "Bactériothérapie intestinale," in *Médicaments microbiens. Bactériothérapie, vaccination, sérothérapie* (Paris: Librairie J.-B. Baillière, 1909), pp. 5–43.

Unlike antibiotics that delete both harmful and beneficial intestinal microbes, probiotics consists of providing live microbes, for example, in foods such as yogurt, to bolster resistance to disease. Gerald W. Tannock, editor, *Probiotics: A Critical Review* (Norfolk, England: Horizon Scientific Press, 1999).

Another current research interest focuses on the intestinal system and the relationships of its microbiota interacting with cells of the immune system and those lining the digestive tract. Vance J. McCracken and Robin G. Lorenz, "The Gastrointestinal Ecosystem: A Precarious Alliance Among Epithelium, Immunity and Microbiota." *Cellular Microbiology* 3 (2001): 1–11.

22. RD, Russell W. Schaedler, and Mallory Stephens, "The Effect of Antibacterial Drugs on the Fecal Flora of Mice," *JEM* 117 (1963): 231–243; RD, Russell W. Schaedler, and Richard L. Costello, "The Effect of Antibacterial Drugs on the Weight of Mice," *JEM* 117 (1963): 245–257; and Dwayne C. Savage and RD, "Alterations in the Mouse Cecum and Its Flora Produced by Antibacterial Drugs," *JEM* 128 (1968): 97–110.

23. RD and Russell W. Schaedler, "The Effect of the Intestinal Flora on the Growth Rate of Mice, and on Their Susceptibility to Experimental Infections," *JEM* 111 (1960): 407–417; idem, "The Susceptibility of Mice to Bacterial Endotoxins," *JEM* 113 (1961): 559–570; RD, Russell W. Schaedler, and Richard Costello, "Composition, Alteration, and Effects of the Intestinal Flora," *Federation Proceedings* 22 (1963): 1322–1329; idem, "Association of Germfree Mice with Bacteria Isolated from Normal Mice," *JEM* 122 (1965): 77–82; and idem, "The Influence of Endotoxin Administration on the Nutritional Requirements of Mice," *JEM* 122 (1965): 1003–1015.

24. Russell W. Schaedler and RD, "The Fecal Flora of Various Strains of Mice. Its Bearing on Their Susceptibility to Endotoxin," *JEM* 115 (1962): 1149–1160; and RD, Dwayne C. Savage, and R.W. Schaedler, "The Indigenous Flora of the Gastrointestinal Tract," *Diseases of the Colon & Rectum* 10 (1967): 23–34.

25. RD and Russell W. Schaedler, "Some Biological Effects of the Digestive Flora," *American Journal of the Medical Sciences* 244 (1962): 265–271; Rose Mushin and RD, "Colonization of the Mouse Intestine with Escherichia coli," *JEM* 122 (1965): 745–757; and idem, "Coliform Bacteria in the Intestine of Mice," *JEM* 123 (1966): 657–663.

26. RD, "The Microbiota of the Gastrointestinal Tract," *Gastroenterology* 51 (1966): 868–874; and RD, Dwayne C. Savage, and Russell W. Schaedler, "Biological Freudianism: Lasting Effects of Early Environmental Influences," *Pediatrics* 38 (1966): 789–800. A recent series of articles reflecting on this 1966 article in *Pediatrics* appeared in the *International Journal of Epidemiology* 34 (2005): 1–20.

27. RD, Savage, and Schaedler, "Biological Freudianism"; and RD, "Biological Remembrance of Things Past," *Bulletin of the Philadelphia Association for Psychoanalysis* 17 (1967): 133–148.

28. RD, "Environmental Determinants of Human Individuality," in *Issues in Human Development: An Inventory of Problems, Unfinished Business and Directions for Research*, edited by Victor C. Vaughan, III (Washington, D.C.: U.S. Government Printing Office, 1967), pp. 5–15.

29. RD, "Man and His Environment: Biomedical Knowledge and Social Action," *Perspectives in Biology and Medicine* 9 (1966): 523–536; Leonardo J. Mata, R.A. Kromal, J.J. Urrutia, and B. Garcia, "Effect of Infection on Food Intake and the Nutritional State: Perspectives as Viewed from the Village," *American Journal of Clinical Nutrition* 30 (1977): 1215–1227; and Leonardo J. Mata, *The Children of Santa Maria Cauque: A Prospective Field Study of Health and Growth* (Cambridge: M.I.T. Press, 1978).

30. RD, Russell W. Schaedler, and Richard Costello, "Lasting Biological Effects of Early Environmental Influences. I. Conditioning of Adult Size by Prenatal and Postnatal Nutrition," *JEM* 127 (1968): 783–799.

31. Egilde Seravalli and RD, "Lasting Biological Effects of Early Environmental Influences. II. Lasting Depression of Weight Caused by Neonatal Contamination,"

JEM 127 (1968): 801–818; and RD, Chi-Jen Lee, and Richard Costello, "Lasting Biological Effects of Early Environmental Influences. V. Viability, Growth, and Longevity," *JEM* 130 (1969): 963–977.

32. Chi-Jen Lee and RD, "Lasting Biological Effects of Early Environmental Influences. VI. Effects of Early Environmental Stresses on Metabolic Activity and Organ Weights," "VII. Metabolism of Adenosine 3′, 5′-monophosphate in Mice Exposed to Early Environmental Stress," and "VIII. Effects of Neonatal Infection, Perinatal Malnutrition, and Crowding on Catecholamine Metabolism of Brain," *JEM* 133 (1971): 147–155; 135 (1972): 220–234; and 136 (1972): 1031–1042.

33. Dennis P. Andrulis, Lisa M. Duchon, and Hailey M. Reid, "Healthy Cities, Healthy Suburbs," the first of five reports, *The Social and Health Landscape of Urban and Suburban America*, sponsored by The Robert Wood Johnson Foundation, August 2002, www.downstate.edu/urbansoc_healthdata.

34. B.J. Stoll et al., "Changes in Pathogens Causing Early-Onset Sepsis in Very-Low-Birth-Weight Infants," *New England Journal of Medicine* 347 (2002): 240–247; RD, Schaedler, and Costello, "Effect of Antibacterial Drugs"; and RD, "Lasting Biological Effects of Early Influences," *Perspectives in Biology and Medicine* 12 (1969): 479–491.

35. Alexander Kessler, "Interplay between Social Ecology and Physiology, Genetics and Population Dynamics of Mice" (Ph.D. dissertation, The Rockefeller University, 1966).

36. John B. Calhoun, "Population Density and Social Pathology," *Scientific American* 206 (1962): 139–148; Paul R. Ehrlich, *The Population Bomb* (New York: Ballantine Books, 1968); and RD, *Man Adapting*, passim.

37. Glenn L. Paulson, "Physiological Stress and Environmental Insults: Toxicity in the Mouse of Long-Term Exposure to Dieldrin and DDT" (Ph.D. dissertation, The Rockefeller University, 1971).

38. RD, [Review of *The Complications of Modern Medical Practice* by David M. Spain], *Archives of Environmental Health* 8 (1964): 769–770.

39. RD, "Biological Sciences and Medicine," in *The Great Ideas Today*, edited by Robert Hutchins and Mortimer Adler (Chicago: Encyclopedia Britannica, 1964), pp. 224–271; and idem, *Man Adapting*, p. 220.

40. In 1960, Margaret Mead and Barry Commoner published a report from the AAAS Committee on Science in the Promotion of Human Welfare: "Science and Human Welfare," *Science* 132 (8 July 1960): 68–73. They were spurred by a concern for the public's need to know and understand scientific issues related to public policy, and their goal was to have its scientist members educate the public on these issues. As a result, several scientists' information committees sprang up around the country, one in St. Louis, where Commoner was then located, and another in New

York at The Rockefeller Institute. In February 1963, these committees formed the nationwide federation called SIPI. Their concerns with air pollution, transportation, automation, and other urban environmental issues were published in the magazine *Environment*, formerly *Scientist and Citizen*. Members of the local chapter at Rockefeller included RD, Theodosius Dobzhansky, Edward Tatum, Jules Hirsch, Ludwig Edelstein, and several graduate students. They met weekly in the Welch Hall lunch room to discuss ways to accumulate and provide information on problems ranging from nuclear waste disposal to lead in paint. RD also contributed several articles and reviews to *Environment* and to SIPI's conferences. Within a few years, he became opposed to their emphasis on negative issues and, in concert with his general rejection of doomsayers, left the organization.

41. "Lead Poisoning is Affecting 112,000 Children Annually, Specialists Report," *The New York Times*, 26 March 1969; and Sandra Blakeslee, "Experts Recommend Measures to Cut Lead Poisoning in Young," *The New York Times*, 27 March 1969.

42. W.J. Rogan et al., "The Effect of Chelation Therapy with Succimer on Neuropsychological Development in Children Exposed to Lead," *New England Journal of Medicine* 344 (2001): 1421–1426.

43. RD, "Toxic Factors in Enzymes Used in Laundry Products," *Science* 173 (1971): 259–260; and Lawrence M. Lichtenstein et al., "Sensitization to the Enzymes in Detergents," *Journal of Allergy* 47 (1971): 53–55.

44. Paul Brodeur, "The Enigmatic Enzyme," *The New Yorker*, 16 January 1971, pp. 68–74.

45. Detlev W. Bronk to RD, 11 March 1960, RD Papers.

46. François Rabelais, *Gargantua and Pantagruel*, Book II, Chapter 8.

47. RD, *The Torch of Life*, p. 19.

48. This topic was discussed in other important lectures, including "Medical Utopias," *Daedalus* 88 (1959): 410–424; and "Medicine as a Social Science," *Arizona Medicine* 16 (1959): 668–673.

49. Some lectures on this topic include "Man Meets his Environment," in *Health and Nutrition. Science, Technology and Development*, volume six, U.S. Papers prepared by the International Development Agency, U.S. State Department, for the United Nations Conference "Application of Science and Technology for the Benefit of the Less Developed Areas" (Washington, D.C.: United States Government Printing Office Document 5096, 1963), pp. 1–11; "Emerging Patterns of Disease," in *Man Under Stress*, edited by Seymour Farber (San Francisco: University of California, 1964), pp. 120–130; "Promises and Hazards of Man's Adaptability," in *Environmental Quality in a Growing Economy*, edited by Henry Jarrett (Baltimore: The Johns Hopkins Press, 1966), pp. 23–43; and "The Pestilence that Stealeth in the Darkness," *Newsletter of the Environmental Mutagen Society* 5 (1971): 9–11.

50. RD, *Man Adapting*, pp. 330–343.

51. RD to Cornelius van Niel, 17 January 1964, RD Papers.

52. RD to H. Kihara, 11 July 1968, RD Papers.

53. David Mechanic, Foreword to Dubos and Dubos, *The White Plague* (New Brunswick: Rutgers University Press, 1987), p. xi. Also, Thomas McKeown, *The Origins of Human Disease* (Oxford, England: Basil Blackwell, 1988); and Anthony J. McMichael, *Human Frontiers, Environments, and Disease: Past Patterns, Uncertain Futures* (Cambridge: University Press, 2001).

In 1966, the U.S. Surgeon General announced a Division of Environmental Health Sciences, which three years later became an independent Institute of the National Institutes of Health. Among its research concerns are environmental exposure to asbestos, lead, hormones, and agricultural pollution and trying to establish biological markers to measure the body's exposure to toxins.

54. RD, "Humanistic Biology," *American Scientist* 53 (1965): 4–19; and idem, "Science and Man's Nature," *Daedalus* 94 (1965): 223–244.

55. Louis Vaczek to RD, 5 January 1979, RD Papers.

56. RD, *Dreams of Reason*, p. 167.

57. Charles P. Snow, *The Two Cultures and the Scientific Revolution* (Cambridge: University Press, 1959).

58. RD, "Biological Limitations of Freedom: Ethical Issues in Genetic Manipulation and Biologic Conditioning," in *Man and Life* (Cincinnati: University of Cincinnati, 1969), pp. 14–28.

59. RD, *So Human an Animal*, p. 120.

60. RD, *Dreams of Reason*, p. 165.

61. William Kieffer, "Panel Discussion with René Dubos, Henry Margenau, and Melvin Calvin," audiotapes for *The Pursuit of Significance. I. The Natural Sciences* (Wooster, Ohio: The College of Wooster Centennial Productions, 19 February 1966), RD Papers, Box 105, AT-3.

62. Jacques Barzun, in the Panel Discussion, in *Man and Life*, pp. 59–86.

63. RD, "Logic and Choices in Science," *Proceedings of the American Philosophical Society* 107 (1963): 365–374.

64. Lewis Mumford to RD, 10 June 1973, RD Papers.

65. During this same period RD also turned down offers demanding organizational and administrative skills from the Pasteur Institute and the Collège de France.

66. Ernst Haeckel, *The History of Creation*, sixth edition, translated by E. Ray Lankester (New York: D. Appleton, 1914), volume two, p. 477. In this translation, based on the 1868 edition, ecology is defined as "correlations between all organisms

which live together in one and the same locality . . . furnished by the theory of the adaptations of organisms to their surroundings."

67. The distinction made by some scientists that ecology belongs to the natural sciences whereas human ecology falls within the social sciences remains an active lumper or splitter issue. Some useful perspectives on the origin, nature, and scope of human ecology appear in these publications: Charles C. Adams, "The Relation of General Ecology to Human Ecology," *Ecology* 16 (1935): 316–335; Amos H. Hawley, *Human Ecology* (New York: Ronald Press Company, 1950); Donald Worcester, *Nature's Economy: A History of Ecological Ideas* (Cambridge: University Press, 1977); Mark J. McDonnell and Steward T.A. Pickett, editors, *Humans as Components of Ecosystems* (New York: Springer-Verlag, 1993); Robert P. McIntosh, *The Background of Ecology: Concept and Theory* (Cambridge: University Press, 1985); Barrington Moore, "The Scope of Ecology," *Ecology* 1 (1920): 3–5; Paul B. Sears, "Human Ecology: A Problem in Synthesis," *Science* 120 (1954): 959–963; and Gerald L. Young, editor, *Origins of Human Ecology* (Stroudsburg, Pennsylvania: Hutchinson Ross Publishing Company, 1983).

68. Lewis Mumford, *The Brown Decades* (New York: Harcourt, Brace and Company, 1931), p. 78.

69. H.G. Wells, Julian S. Huxley, and George P. Wells, *The Science of Life* (New York: Doubleday, Doran and Company, 1935), pp. 961–962; and H.G. Wells, *The Work, Wealth and Happiness of Mankind* (New York: Doubleday, Doran and Company, 1931), p. 35.

70. Paul B. Sears, *This Is Our World* (Norman: University of Oklahoma Press, 1937).

71. Paul B. Sears, *The Ecology of Man* (Eugene: Oregon State System of Higher Education, 1957), p. 24; and Paul Shepard, "Whatever Happened to Human Ecology?" *BioScience* 17 (1967): 891–894.

72. Roderick Nash, *Wilderness and the American Mind*, third edition (New Haven: Yale University Press, 1982); and idem, *The Rights of Nature: A History of Environmental Ethics* (Madison: University of Wisconsin Press, 1989).

73. RD, *A God Within*, p. 45.

74. RD to Gardner Stout, President of the American Museum of Natural History, 19 May 1971, RD Papers.

75. Max Oelschlaeger, *The Idea of Wilderness: From Prehistory to the Age of Ecology* (New Haven: Yale University Press, 1991), pp. 292–301.

76. RD, "Human Ecology," in *Fifth World Congress of Gynaecology and Obstetrics* (Sydney, Australia: Butterworths, 1967), pp. 15–23; idem, "Surroundings Shape Man and His Progress," *Journal of Environmental Health* 31 (1969): 446–452; and idem, "The Crisis of Man in His Environment," in *Health and the Social Environment*,

edited by Paul M. Insel and Rudolf H. Moos (Lexington, Massaschusetts: Lexington Books, 1974), pp. 361–366.

77. Paul B. Sears, "Ecology—A Subversive Subject," *BioScience* 14 (1964): 11–13; and Paul Shepard and Daniel McKinley, editors, *The Subversive Science: Essays Toward an Ecology of Man* (Boston: Houghton Mifflin Company, 1969).

78. RD, "Human Ecology," *WHO Chronicle* 23 (1969): 499–504.

79. RD, "The Human Environment," *Science Journal* (London) 5A (1969): 75–80.

80. RD, *Reason Awake*, p. 173.

81. Paul B. Sears, "Prelude to Harmony," *The American Scholar* 39 (1969–1970): 722–724.

82. Young, *Origins of Human Ecology*, p. 2.

83. Shosuke Suzuki, Richard J. Borden, and Luc Hens, editors, *Human Ecology—Coming of Age: An International Overview* (Brussels, Belgium: Vrije Universiteit Brussel Press, 1991).

84. Peter Kihss, "Dr. Dubos Wins 'American Nobel,'" *The New York Times*, 5 October 1966.

85. RD, *So Human an Animal*, p. 192.

86. This theme was also presented in several lectures about this time, including "The Biosphere: A Delicate Balance Between Man and Nature," *UNESCO Courier* 22 (1969): 6–15; *A Theology of the Earth* (Washington, D.C.: Smithsonian Institution, 1969); and "Life as an Endless Give and Take with Earth and All Her Creatures," *Smithsonian* 1 (1970): 8–17 (this was the lead article in the premier issue).

87. RD, *Man Adapting*, p. 357.

88. Lewis Thomas, "Medicine and Health," 10 December 1982, RD Papers, Box 1, Folder 7.

89. François Lanthenas, *De l'influence de la liberté sur la santé, la morale, et le bonheur* (Paris: Cercle Social, 1792), p. 18; quoted by Michel Foucault, *The Birth of the Clinic; An Archaeology of Medical Perception*, translated by A.M. Sheridan Smith (New York: Pantheon Books, 1973), p. 35.

90. RD, "Logic and Choices in Science," p. 367.

Chapter 6 Health as Creative Adaptation

1. John Culhane, "*En Garde* Pessimists! Enter René Dubos," *The New York Times Magazine*, 17 October 1971, pp. 44–64, 68.

2. "Medicine's Living History: Dr. René Dubos," *Medical World News* 16 (5 May 1975): 77–87.

3. RD, *The Professor*, p. 31.

4. RD to Henriette Noufflard, 18 July 1958, author's papers.

5. Sophie Lannes, "Une médecine si humaine," *L'Express*, 3 November 1979, pp. 92–104.

6. Rae Goodell, *The Visible Scientists* (Boston: Little, Brown, 1977), p. 32.

7. RD to Melvin M. Tumin, 16 November 1970, RD Papers.

8. RD to W. Hoffman, 3 November 1975, RD Papers.

9. RD, *Genius of the Place*.

10. This lecture information was compiled for the author by Jean Porter Dubos from her personal diary.

11. Dubos' other activities reveal an even larger constellation of publics. Of his two dozen full-length books, every one was translated into at least one foreign language, with an average of five translations; one appeared in nineteen languages. After 1965, during the last years of his life, he was a visible scientist who was the subject of at least one hundred newspaper and magazine articles and interviews. He was featured in a half-dozen full-length television documentaries; interviewed for more than 150 television and radio programs in the United States, Europe, Japan, and Australia; and received twenty-eight of his forty-one honorary degrees and a dozen major international awards.

12. Through the gracious, generous persistence of Laurence Rockefeller, in 1969 President Richard Nixon appointed Dubos a member of the Citizens' Advisory Committee on Environmental Quality. Though he served for six years, and attended its meetings, he felt this appointment was inconsistent with his nature and aversion to engaging in politics, advocacy matters, and any form of activism. In committee meetings, however, he was known to speak out about the quality of human life in the environment rather than the quality of the environment itself.

13. Gerard Piel, Foreword to *The World of René Dubos: A Collection from His Writings*, edited by Piel and Osborn Segerberg, Jr. (New York: Henry Holt and Company, 1990), p. xxiv.

14. RD, DO 44 (1974–1975): 8–13; idem, DO 44 (1975): 174–180; idem, "Genetic Engineering," an op-ed article, *The New York Times*, 21 April 1977; idem, "L'energie verte et la decentralization," *Critère* 23 (1978): 259–268; and idem, "Energy Galore," University of Washington, Seattle, 1978, 30 typescript pages, RD Papers, Box 29, Folder 21.

15. Richard Halloran, "Nader to Press for GM Reform," *The New York Times*, 8 February 1970.

16. RD to Frederick Seitz, 13 April 1970, RD Papers.

17. "May 11: National & Campus Issues Spark Full Day of Activities," *News and Notes* 1 (18 May 1970): 1–2.

18. RD to John Adams, 12 Dec 1981, RD Papers.

19. RD, "The Socialization of Troubled Teenagers, or the Wasted Potential of Adolescents," January 1982, 6 typescript pages, RD Papers, Box 32, Folder 4. Speaking as a biologist, RD often faulted society for not recognizing that teenagers with active minds need satisfying, worthwhile forms of expression if they are to become creative, healthy adults. Otherwise, he predicted, their lives would give way to sexual impulses, juvenile delinquencies, drug addictions, or philosophies of despair.

20. Among these organizations were the United Nations, Gateway National Park Planning Commission, Natural Resources Defense Council, The Parks Council, New York Citizens for Balanced Transportation that became the Westway Advisory Committee, The Council on the Environment for New York City, Scenic Hudson Preservation Conference, the Regional Plan Association, and Scientists' Institute for Public Information.

21. RD to E.D. Kilbourne, Autumn 1971, RD Papers. Noteworthy in the correspondence of Dubos during the 1970s are the overwhelming (nearly three-fourths of letters) refusals and regrets to accept lecture engagements. His basis for accepting invitations, however, seems arbitrary. There is a clear favoritism toward small liberal arts colleges, exclusive lectures, and small, interactive seminars with students and faculty and a clear bias against large international symposia on grandiose topics having many other speakers.

22. RD, [Review of *The Closing Circle* by Barry Commoner], *Environment* 14 (1972): 48–49.

23. RD, DO 40 (1970–1971): 16–20; and idem, "Trend is Not Destiny: An Ecologist's View," in *Children and the Environment: Lucy Sprague Mitchell Memorial Conference* (New York: Bank Street College of Education, 1971), pp. 12–24.

24. The quotation appears in various places, including *Newsweek*, 28 May 1979, pp. 85–88.

25. Joseph Kastner, "The 'Miracle' on Jamaica Bay Didn't Happen Overnight," *Smithsonian* 21 (July 1990): 110–116.

26. David Bird, "Though Still Polluted, Jamaica Bay is Reviving," *The New York Times*, 10 September 1970.

27. Truman Temple, "Think Globally, Act Locally: An Interview with Dr. René Dubos," *EPA Journal*, April 1978, pp. 4–11.

28. RD, DO 44 (1975): 528–531; idem, "Restoring a Treasure," an op-ed article, *The New York Times*, 23 July 1980; and "Fourteen on Panel to Consult on Park for Westway," *The New York Times*, 21 June 1981.

29. RD to John Adams, 18 December 1981, RD Papers.

30. Garry Pierre-Pierre, "After Two Decades, Work Begins on Far Less Ambitious Westway," *The New York Times*, 2 April 1996.

31. RD to E. Menert, 26 October 1979, RD Papers.

32. John G. Mitchell, "A Perfect Day for Earth," *Audubon*, March 1990, pp. 109–123.

33. David Bird, "Long Battle Seen to End Pollution," *The New York Times*, 1 February 1970.

34. RD, "The Environmental Teach-In," and "The Limits of Adaptability," in *The Environmental Handbook*, edited by Garrett De Bell (New York: Ballantine/Friends of the Earth, 1970), pp. 8–9 and 27–30.

35. Frank Herbert, editor, *New World or No World* (New York: Ace Books, 1970).

36. RD, "An Earth Day Talk," *The Morton Arboretum Quarterly* 6 (1970): 1–4.

37. RD, "Mere Survival is Not Enough for Man," *Life* 69 (24 July 1970): 2; and idem, "Why Survival is Not Enough," *Reader's Digest* 97 (October 1970): 111–112.

38. RD, "The Predicament of Man," *Science Policy News* (London) 2 (May 1971): 64–69; and idem, *Genius of the Place*, p. 1.

39. Some important lectures to these audiences include "Man and His Environment: Scenarios for the Future," *New York University Educational Quarterly* 2 (1971): 2–7; "Man-Made Environments," *The Journal of School Health* 41 (1971): 339–343; "Shelters, Open Spaces and Adaptations," Richard Neutra Lecture, 25 October 1972, 35 typescript pages, RD Papers, Box 28, Folder 31; "In Praise of Diversity," *American Institute of Architects Journal* 58 (1972): 30–33; "Biological Determinants of Urban Design," *The Canadian Architect* 18 (May 1973): 41–44; "Sensory Perception and the Museum Experience," *Museum News*, October 1973, pp. 50–51; and "Homeostasis, Illness, and Biological Creativity," *Lahey Clinic Foundation Bulletin* 23 (1974): 94–100.

40. RD, "Environment," in *Dictionary of the History of Ideas: Studies of Selected Pivotal Ideas*, edited by Philip P. Wiener (New York: Scribner, 1973–1974), volume 2, pp. 120–127.

41. Lannes, "Une médecine si humaine."

42. Stephen Jay Gould, "A Tale of Two Worksites," *Natural History*, October 1997, pp. 18–22, 29, 62–68.

43. Salvador E. Luria, *Life: The Unfinished Experiment* (New York: Scribner, 1973).

44. RD to S. Luria, 17 October 1972, RD Papers.

45. RD, "Credo of a Biologist," *Journal of Religion and Health* 10 (1971): 313–323; and idem, *A God Within*, p. 249.

46. RD, "Environment"; and idem, "Experimental Medicine Revisited: Claude Bernard after 100 Years," Harvard Medical School Interdisciplinary Seminar, 11 January 1968, 16 typescript pages, RD Papers, Box 27, Folder 14.

47. RD, "Homeostasis"; RD also objected to the Gaia concept as one of homeostasis and favored instead the concept of Earth as a living, evolving organism: RD, "Gaia and Creative Evolution," *Nature* 282 (1979): 154–155.

48. RD, *So Human an Animal*, p. 236.

49. Stephen Boyden to the author, 2 August 1999.
The steady-state theory is currently coming under attack from ecologists who are finding evidence of cyclical occurrences of pollen deposits, tree-ring widths, and animal populations over time. According to Daniel Botkin, scientists now believe this theory is "wrong at the levels of population and ecosystems. Change now appears to be intrinsic and natural at many scales of time and space in the biosphere." Daniel Botkin, *Discordant Harmonies: A New Ecology for the Twenty-first Century* (New York: Oxford University Press, 1990), p. 10.

50. Dubos probably did not know *homeokinesis* was used by August Weismann in 1893 to describe cellular division in which two daughter nuclei receive chromosomes of the same kind: *The Germ-Plasm: A Theory of Heredity*, translated by W.N. Parker and H. Rönnfeldt (New York: Scribner, 1893), p. 34.

51 "U.N. Enlists Dubos for Pollution Study," *The New York Times*, 29 April 1971.

52. RD to Barbara Ward, 28 September 1971, RD Papers.

53. According to Lord Solly Zuckerman, a participant in the 1972 Stockholm Conference and the 1992 United Nations Conference on Environment and Development in Rio de Janeiro, *Only One Earth* was "as relevant [in 1992] as it was when it appeared" for providing a basic understanding of all environmental problems. Lord Zuckerman, "Between Stockholm and Rio," *Nature* 358 (1992): 273–276.

54. RD, "Unity through Diversity," in *Who Speaks for Earth?*, edited by Maurice Strong (New York: W.W. Norton, 1973), pp. 33–42.

55. RD to Joanna Underwood, 11 July 1975, RD Papers.

56. RD to Paul Ehrlich, 18 October 1977, RD Papers.

57. RD, DO 46 (1977): 152–158.

58. RD, "The New Frontiers in American Society," Princeton University, 8 February 1977, 24 typescript pages, RD Papers, Box 29, Folder 2.

59. Temple, "Think Globally, Act Locally."

60. *Wall Street Journal*, 17 May 1979, pp. 17–19; and *Newsweek*, 28 May 1979, pp. 85–88. At the time this advertisement appeared, SmithKline was considering support for a television documentary based on RD's life. When this project failed, RD

integrated the script he had composed into his final book *Celebrations of Life*. The script exists as "A Celebration of Life," August 1979, 127 typescript pages, RD Papers, Box 29, Folder 19 and Box 32, Folder 8.

61. Essayist and moral critic Wendell Berry has criticized "Think Globally, Act Locally" as a perilous slogan. Actually, RD's reservations about global thinking are as cautious as those Berry advocated. Wendell Berry, "Out of Your Car, Off Your Horse," in *Sex, Economy, Freedom, and Community* (New York: Pantheon Books, 1993), pp. 19–26.

62. RD, *Wooing of Earth*, pp. 156–157.

63. RD, *Celebrations of Life*, p. 125.

64. George Perkins Marsh, *Man and Nature, or Physical Geography as Modified by Human Action*, edited by David Lowenthal (Cambridge: Belknap Press of Harvard University Press, 1967), p. 381; and David Lowenthal, *George Perkins Marsh: Prophet of Conservation* (Seattle: University of Washington Press, 2000).

65. Theodore Roosevelt, "Opening Address by the President," May 1908, in *Proceedings of a Conference of Governors in the White House*, edited by Newton C. Blanchard (Washington, D.C.: United States Government Printing Office, 1909).

66. In addition to Burroughs, Marsh, and Roosevelt, RD was greatly influenced on this topic by his study of these texts: William L. Thomas, Jr., editor, *Man's Role in Changing the Face of the Earth* (Chicago: University of Chicago Press, 1956); Bernard Rudofsky, *Architecture Without Architects* (Garden City, New York: Doubleday, 1964); Clarence J. Glacken, *Traces on the Rhodian Shore* (Berkeley: University of California Press, 1967); Paul Shepard, *Man in the Landscape: A Historic View of the Esthetics of Nature* (New York: Knopf, 1967); Nan Fairbrother, *New Lives, New Landscapes* (New York: Knopf, 1970); Yi-Fu Tuan, *Topophilia: A Study of Environmental Perception, Attitudes, and Values* (Englewood Cliffs, New Jersey: Prentice-Hall, Inc., 1974); and Jay Appleton, *The Experience of Landscape* (New York: Wiley, 1975). RD also relished every issue of the journal *Landscape*, founded by John Brinckerhoff Jackson in 1951.

67. RD, *Louis Pasteur*, p. 391.

68. RD, untitled speech at College of the Atlantic, Bar Harbor, Maine, 1 February 1978, enclosed with a letter from Paul Beltramini to RD, 28 March 1978, 2 typescript pages, RD Papers.

69. RD, *Humanizing the Earth*; and idem, "Human Touch Often Improves the Land," *Smithsonian* 3 (December 1972): 18–29.

70. Other aspects of "improving on nature" appear in these texts by RD: "The Biological Basis of Urban Design,"*Anthropopolis: City for Human Development*, edited by C.A. Doxiadis (New York: Norton, 1975); "Humanizing the Earth," *Science* 179 (1973): 769–772; "Man-Made Environments," in *Environmental Spectrum: Social*

and Economic Views on the Quality of Life, edited by Ronald O. Clarke and Peter C. List (New York: D. Van Nostrum, 1974), pp. 96–110; "This is the Place—A Man-Made Place," Vermont Legislature, January 1974, 22 typescript pages, RD Papers, Box 28, Folders 14 and 32; "The Necessary Pleasures of Park Life," Architectural League of New York, 12 June 1978, 4 manuscript and 6 typescript pages, RD Papers Box 29, Folder 12; "Le mythe de la nature," in *Congrès national de la protection des plantes* (Boulogne-Billancourt, France: Union des Industries de la Protection des Plantes, 1980), pp. 41–61; "Beyond the Garden Wall," *The Sciences* 22 (1982): 10–14; and "Shelters—Their Environmental Conditioning and Social Relevance," in *Shelter: Models of Native Ingenuity*, edited by James Marston Fitch (Katonah, New York: The Gallery, 1982), pp. 9–15.

71. Beth Ann Krier, "A Maverick Award-Winning Ecologist," *Los Angeles Times*, 15 April 1976; RD to J.E. McKee, 19 September 1975, RD Papers; and RD, "Symbiosis Between the Earth and Humankind," *Science* 193 (1976): 459–462. RD shared the Tyler Prize with Charles S. Elton and Abel Wolman.

72. Paul Ehrlich to RD, 23 May 1978, RD Papers.

73. Leo Marx to Ruth Eblen, 7 September 1978, RD Papers.

74. RD to Leo Marx, 10 October 1978; and Leo Marx to RD, 23 October 1978, RD Papers.

75. Roderick Nash, *Wilderness and the American Mind*, third edition (New Haven: Yale University Press, 1982), pp. 241–244, 380, 383–384.

76. RD, "Symbiosis."

77. RD and Edward Abbey with photographs by Ernst Haas, "The Eighth Day," *Geo Magazine,* July 1979, pp. 91–116.

78. William K. Reilly, New York City, 10 December 1982, RD Papers, Box 1, Folder 7.

79. RD, "Wilderness and Humanized Nature in America," Aldo Leopold Memorial Lecture, Madison, Wisconsin, 21 June 1978, 24 typescript pages, RD Papers, Box 29, Folder 15.

80. John E. Ross to the author, 29 November 1997.

81. RD, Untitled manuscript on presenting the Lincoln Land Policy Award to Lewis Mumford, Cambridge, Massachusetts, 18 October 1979, 2 typescript pages, Box 30, Folder 9.

82. RD, "Wilderness," p. 22.

83. Aldo Leopold, *A Sand County Almanac* (New York: Oxford University Press, 1949), p. 224.

84. Quoted in Lowenthal, *Marsh*, p. 411.

85. RD, "The 5 E's of Environmental Management," in Piel, *The World of René Dubos*, pp. 400–408.

86. *Wooing of Earth* was nominated for the 1980 American (now called National) Book Award.

87. Krier, "A Maverick."

88. RD, "A Family of Landscapes," circa 1974, 58 typescript pages, RD Papers, Box 32, Folder 9 and Box 30, Folder 18.

89. RD to Charles Scribner, 22 and 29 June 1979, RD Papers.

90. Rabindranath Tagore, "A Poet's School," in *Towards Universal Man* (New York: Asia Publishing House, 1961), p. 294.

91. Valéry Giscard d'Estaing, "L'écologie objectif de civilisation," *Environnement et Cadre de Vie* 13 (December 1980): 6–11. Translated from the original French.

In 1975 a René Dubos Forum was formed by educators as an outgrowth of their secondary school environmental program. His personal participation in its few symposia was comparatively modest and completely independent of his full schedule of lectures and writing at Rockefeller. In 1981, about a year before he died, the forum was reorganized into the René Dubos Center for Human Environments. RD thought it should have "écologie civilisatrice" as its mission and that it would collect case studies and create a library documenting the creative aspects of human interventions into nature. Jean Dubos and others recounted that he soon became disillusioned with the center and just before he died was considering how to dissociate himself from it. Jean Dubos, Lucille Porter, Henriette Noufflard, and Cynthia Chase, personal communications to the author.

92. Leopold, *Sand County Almanac*, pp. 195–196.

93. Aldo Leopold, "The Farmer as a Conservationist," *American Forests* 45 (1939): 294–299, 316, 323.

94. RD, *The Unseen World*, p. 101.

95. RD and Alex Kessler, "Integrative and Disintegrative Factors in Symbiotic Associations," in *Symbiotic Associations*, edited by P.S. Nutman and Barbara Mosse (London: Cambridge University Press, 1963), pp. 1–11; and RD, "The Lichen Sermon," in Piel, *The World of René Dubos*, pp. 409–415.

96. Maurice Caullery attributes the first use of symbiosis to Heinrich Anton de Bary, *Die Erscheinung der Symbiose* (Strassburg: Cassel, 1879); quoted in Caullery, *Parasitism and Symbiosis*, translated by Averil M. Lysaght (London: Sidgwick and Jackson, 1952), p. 217; see also Jan Sapp, *Evolution by Association: A History of Symbiosis* (New York: Oxford University Press, 1994).

97. Henry David Thoreau, *Journals*, 17 February 1859.

98. RD, *Beast or Angel*, p. 185; and RD, *Celebrations of Life*, p. 125.

99. Albert Blanchard to RD, 16 January 1973, RD Papers. Translated from the original French.

100. RD, "Creative Adaptations," *Digest of Neurology and Psychiatry* 40 (1972): 341.

101. RD, *Beast or Angel*, pp. 209–210.

102. RD to E.O. Wilson, 9 June 1978, RD Papers; and RD, *Celebrations of Life*, pp. 5, 13.

103. Theodosius Dobzhansky, [Review of *Beast or Angel*], *The Quarterly Review of Biology* 50 (1975): 514–515.

104. RD and Jean-Paul Escande, *Quest: Reflections on Medicine, Science and Humanity*, translated by Patricia Ranum (New York: Harcourt Brace Jovanovich, 1980). The original French edition was published as *Chercher. Des médecins, des chercheurs . . . et des hommes* (Paris: Éditions Stock, 1979). See also Robert Serrou, "Un grand savant français parle/René Dubos," *Paris Match*, 4 April 1980, pp. 3–10.

105. RD, "Baghdad on the Hudson," an op-ed article, *The New York Times*, 22 May 1975.

106. RD, DO 46 (Autumn 1977): 424–430; idem, "Creative Adaptations to the Future," in *Aspects of American Liberty: Philosophical, Historical, Political*, edited by George Corner (Philadelphia: Memoirs of The American Philosophical Society 118, 1977), pp. 162–173; and idem, "First Person Singular: A Celebration of Life," *United Magazine* 26 (April 1982): 9, 102.

107. RD, *The Professor*, p. 33.

108. RD, DO 46 (Winter 1976–1977): 10–18.

109. RD and Escande, *Quest*, p. 14.

110. RD, "Health as Ability to Function," in *Reflections of America: Commemorating the Statistical Abstract Centennial (1879–1979)*, edited by Norman Cousins (Washington, D.C.: U.S. Department of Commerce, Bureau of the Census, December 1980), pp. 105–112.

111. RD, "Self-Healing: A Personal History," in *Healing Brain; A Scientific Reader*, edited by Robert Ornstein and Charles Swencionis (New York: The Guilford Press, 1990), pp. 135–146.

112. Robert E. Ornstein, and David Sobel, *The Healing Brain* (New York: Simon & Schuster, 1987); Sherwin B. Nuland, *How We Die: Reflections on Life's Final Chapter* (New York: Knopf, 1994); Daniel Callahan, *False Hopes: Why America's Quest for Perfect Health is a Recipe for Failure* (New York: Simon & Schuster, 1998); and

Abraham Verghese, "The Healing Paradox," *The New York Times Magazine*, 8 December 2002.

113. RD, "Health as Ability to Function."

114. Lannes, "Une médecine si humaine."

115. Arthur S. Freese, "Dubos at 80," *Modern Maturity*, August/September 1981, pp. 34–36.

116. RD, "Social Adaptations in Medicine: From the Bedside to the Laboratory," in *Patterns for Progress from the Sciences to Medicine*, edited by John A. Hogg and Jacob C. Stucki (Miami, Florida: Symposia Specialists, 1977), pp. 133–147; idem, "Health and Creative Adaptation," *Human Nature* 1 (1978): 74–82; and idem, "Biological Memory, Creative Associations and the Living Earth," in *The Nature of Life*, edited by William H. Heidcamp (Baltimore: University Park Press, 1978), pp. 1–21.

117. Geneviève Moll and Robert Leblanc. "René Dubos. La médecine ne peut toujours pas tout," *Tempo Médical* 76 (March 1981): 91–94.

118. RD celebrated the human capacity for self-recovery and longevity in a long introduction to Norman Cousins' book *Anatomy of an Illness as Perceived by the Patient* (New York: Norton, 1979). Both men maintained these goals could be achieved with natural biological and psychological responses so that most threats do not result in disease, and they urged physicians and patients to manage and protect these natural drives. The book is still in print.

119. Henri Bergson, *Creative Evolution*, translated by Arthur Mitchell (New York: Henry Holt, 1911), p. xiv.

120. Alfred North Whitehead, *Science and the Modern World* (New York: Macmillan, 1925), pp. 287, 292.

121. RD, *The Torch of Life*, p. 99.

122. RD, "Life-Lessons from the New Biology," 12 April 1981, 31 typescript pages, RD Papers, Box 31, Folder 24.

123. RD to H.J. Kaplan, editor of *Geo Magazine*, n.d., RD Papers.

124. RD, "Life-Lessons," pp. 28, 26.

125. RD, *So Human an Animal*, p. 232.

126. RD, DO 43 (Winter 1973–1974): 10–16.

127. RD, interview by Claire Warga, *Omni* 2 (December 1979): 87, 88, 126, 128.

Index